A Natural History of Families

A Natural History of Families

Scott Forbes

PRINCETON UNIVERSITY PRESS

PRINCETON AND OXFORD

Third printing, and first paperback printing, 2007
Paperback ISBN-13: 978-0-691-13035-4
Paperback ISBN-10: 0-691-13035-3

The Library of Congress has cataloged the cloth edition of this book as follows

Forbes, Scott, 1958–
A natural history of families / Scott Forbes.
p.   cm.
Includes bibliographical references and index.
ISBN 0-691-09482-9 (alk. paper)
1. Reproduction. 2. Reproduction—Social aspects. 3. Family.
4. Parental behavior in animals. I. Title.
QP251.F644 2005
306.87—dc22   2004022425

British Library Cataloging-in-Publication Data is available

This book has been composed in Sabon

Printed on acid-free paper. ∞

pup.princeton.edu

Printed in the United States of America

10 9 8 7 6 5 4 3

Contents

Preface

THIS BOOK has a long genesis, dating back two decades from when I first encountered sibling bullies in ospreys. I owe a debt of gratitude to Spencer Sealy for inspiring my interest in birds; Ron Ydenberg, who taught me how to think like an evolutionary biologist; and Douglas Mock for showing me how much fun it is to be an evolutionary biologist. I thank Sam Elworthy for the invitation to write this book. The Natural Sciences and Engineering Research Council of Canada funded much of my work, for which I am grateful. And of course I must thank my own family, Fraser and Fionnlugh and Maggie, for their unconditional love and support.

A Natural History of Families

Chapter 1

Blame Parents

> *You see a meadow rich in flower & foliage and your memory*
> *rests upon it as an image of peaceful beauty. It is a delusion.*
> *... Not a bird that twitters but is either slayer or [slain*
> *and] ... not a moment passes in that a holocaust, in every*
> *hedge & every copse battle murder & sudden death are the*
> *order of the day.*
>
> —Thomas Henry Huxley

ON SEPTEMBER 25, 1994, Lele, a giant panda at the Beijing Zoo, gave birth to twins. Pandas are critically endangered in their native habitat—the bamboo forests of western China—and any addition to their number is welcome. But the joyous mood was dampened by the dark side of panda parental care. Lele's babies were born 23 minutes apart, the first weighing 5.7 ounces, the second just 1.8 ounces. Almost immediately she abandoned the larger infant in favor of her smaller second-born. Ironically, the name Lele means "double happiness," hardly the case here. Such behavior is not unusual for pandas. Following multiple births mothers normally abandon all but one cub, leaving the others to die. The obvious question is, why? Surely a mother a thousandfold larger than her newborn offspring could easily nourish her tiny twins. But she does not. This sinister habit is not peculiar to giant pandas. Parents commonly play favorites among their progeny, with often lethal consequences.

Children are right: parents are to blame for their woes. Parents are the architects of sibling strife and of conflict between parents and offspring. They make too many babies and then do not provide for all. They play favorites among their children and not only tolerate but foster rivalries. They prefer their offspring to be different, which only makes matters worse. Parents build amphitheaters to observe the contests, and some go so far as to deliberately place offspring gladiators in arenas of sibling warfare from which only one escapes alive. The parents watch from above while their progeny fight a short, bloody, and one-sided battle, and then turn thumbs down on the outcome. But parents are also nurturing and loving. A mother crocodile will gently cradle her babies in fearsome jaws to carry them from their nest to a nearby stream. A male blackbird will risk his own life to defend his nestlings against predators many times his size. Parents that provide the vital resources to sustain and protect their offspring can also be manipulating and cruel.

Parents do not share the view of their children. Parents, in their own view, are self-sacrificing, overtaxed, and underrewarded. To them, the offspring—those objects of parental affection and recipients of lavish and expensive investment—can be too demanding, and too selfish. They are endlessly squabbling among themselves over how their parents' offerings are to be shared, and ever pleading for more.

Vulcan was the celestial artisan of Roman mythology, crafting trinkets for the gods and suits of impenetrable armor for warriors. He also had the misfortune to be born lame. This accident of birth so displeased his mother Juno that she flung him from heaven. Malformed infants were similarly treated in ancient Sparta, where they were cast into chasms called Apothetae, the place of throwaways. And inside the bodies of women everywhere across the globe, embryos with chromosomal defects are quietly discarded, early and out of sight, often without a mother knowing that for a brief time she was pregnant.

Infanticide of newborn daughters was also common in ancient Greece and Rome, as it still is in rural societies all over the world. The Scottish philosopher David Hume once remarked that to kill one's own child is shocking to nature. But is it? Infanticide is not the exclusive province of humans but is surprisingly common in nature. Marsh-nesting blackbirds, for example, treat sons and daughters differently. The male nestlings, being larger and more expensive to nourish, fall victim to starvation faster than females, especially when food is short. Here we might be tempted to suggest that the death of an offspring is attributable to parental neglect, a failure to make adequate provision for all the offspring. But even that view would be too narrow. We need to step back and examine what underpins this dark behavior. And it is this: infanticide, overt or otherwise, is part of a parental conspiracy. Parents produce more progeny than they can normally rear and then render them unequal. Among hyena pups and eaglet broods, fierce and lethal battles rage between brothers and sisters. Beetles and owls dine on their own progeny; hawks, sharks, tadpoles, and snails eat siblings; and aphids even consume their mothers from the inside out. Look just beneath the surface, and you will see Huxley's vision of nature, where chaos and mayhem are ever present.

Baby birds and infant humans would seem at first glance utterly different. One hatches after 12 days in an egg; the other is born after 38 weeks' residence in a womb. One is fed a diet of worms or insects. The other draws sustenance from its mother's blood before birth, and mother's milk after. The differences are many and obvious, yet there are important similarities too. Broods of begging nestlings, with necks stretched upward, beaks agape, stubby wings flapping, and a chorus of chirps, are a familiar sight. They are calling to their parents, and the message is easily decoded: "I am hungry—feed me—bring more!" Sometimes the parents will agree, and hurry away to gather worms or damselflies or other insect morsels to ensure that their nestlings are healthy and well fed. But sometimes not. The baby birds go hungry, often with fatal consequences.

The journey of a human egg from the mother's ovary to the site of implantation occurs out of sight. Over the span of about a week, the egg travels leisurely toward its destination, a blood-engorged uterine wall it nestles against and, if fertilized, grows into. And once there, the little embryo also sends a message to its mother: "I am here—I am healthy—don't forsake me!" Only this time the message is chemical, not acoustic. Sometimes mother will agree, and continue to keep her uterine wall in place and blood engorged so the embryo can draw life-sustaining nutrients and oxygen. But sometimes not. Mother withdraws her life-support system, with lethal consequences.

Do the Good Die Young?

The gaudy colors of male guppies that make them popular with aquarium enthusiasts also make them conspicuous to predators. So do the long tails of male peacocks and widowbirds. Charles Darwin—coauthor with Alfred Russel Wallace of the theory of natural selection—was puzzled by such traits that surrendered personal survival. For what end? These were antithetical to the theory of natural selection. His solution was to propose a theory of sexual selection: traits that impaired survival could nevertheless be selected if they enhanced reproductive success, and gaudy colors and long tails do exactly that by attracting more females. Darwin's new theory did not gain rapid acceptance—even Wallace remained a skeptic. The theory of sexual selection slumbered for a full century until it was resurrected by animal behaviorists in the early 1970s, and it now occupies a central place within the fields of behavioral and evolutionary ecology.

A powerful reformation of how we view families is also now at work, and as with the theory of sexual selection, its roots are deep in time. With somewhat less fanfare than Darwin, the botanist John Buchholz announced a theory of developmental selection in 1922. He suggested that natural selection—because it is subject to chance events was erratic and often weak (it is). Darwin had emphasized the struggle for existence in his description of natural selection, but Buchholz noted that similar processes occur during the embryonic development of plants. He further suggested that parents could (and do) supplement the clumsy process of natural selection with a contest of their own cruel design: it matches offspring in an arena of sibling competition. The winners escape the arena with the reward of continued parental succor. The losers forfeit their lives. Such a system begins at fertilization and ends when offspring leave their parents' care. It thus precedes the period of more familiar natural or "survival" selection, and it follows the process of sexual selection, which occurs when potential parents turn their attention to mating.

During the process of developmental selection, Buchholz surmised, defects in progeny would be exposed quickly and culled swiftly and coldly, avoiding needless waste of precious parental investment in progeny doomed to fail. This dark vision had manipulative parents arranging lethal battles among their own progeny for selfish ends—a foul union of Hobbes's war of all against all with the ruthless efficiency of a Brave New World. And everywhere the botanist Buchholz looked, he could see evidence of such parental machinations. Almost all higher plants employ developmental selection during the process of sexual reproduction. In cycads, ginkgoes, and conifers, multiple embryos engage in an intense life-and-death competition during their development. Weaker individuals are aborted in the earlier stages, and only one reaches its full term to become the seed embryo. But the process Buchholz described is not exclusive to plants. It is widespread in animals too, particularly those that provide their offspring with elaborate and expensive parental care. In fish, amphibians, birds, and mammals, internecine struggles are routine, and it is not the good that die young, but rather precisely the opposite.

The logic of developmental selection as it applies to both plants and animals is simple. Screen progeny early, identify those worthy of continued investment, and discard the rest. Sibling competition can provide such a test, revealing innate defects before parents make an expensive mistake. But as with sexual selection, the traits needed for victory in the sibling arena could prove inimical to their bearers, or to their closest family members. One's closest kin are, after all, potential allies in the larger struggle. Why would parents encourage siblings to kill siblings?

The Family Myth

A mythology has arisen in popular culture about how families are intended to work. In humans they are expected to be warm, nurturing, and loving, and we are shocked to see otherwise. Deviations from this ideal structure appear to be just that, deviations, so much so that a family in conflict is viewed as dysfunctional. The neglect or abuse of children is both surprising and abhorrent: devoted parents are what we expect. Expectation shapes observation, and biologists, like painters, long described what they *saw*, not what was actually there. Infanticide was not a topic of serious scientific interest until the 1970s. Since biologists did not expect it, they did not look for it, and when it was reported, infanticide was considered an aberration caused by stress or a pathology, or was just plain inexplicable. As Einstein once remarked, it is theory that decides what we can observe, and in the 1970s what biologists could observe was changing. Across the world, entomologists, primatologists, mammologists, behavioral ecologists, herpetologists, sociobiologists, and anthropologists in the field and lab started to see the previously unseen. They saw infanticide in the Hanuman langurs and African lions. They saw cannibalism in spiders and scorpions, and sibs killing sibs in herons and egrets, pelicans and boobies, while parents stood by seemingly unconcerned. And in the lab they saw filial infanticide (parents killing their biological offspring) and cannibalism in gerbils and hamsters, and mice. Biologists had seen the unexpected, and they asked the obvious question: why?

Long ago Charles Darwin wrote to his colleague, the botanist Joseph Hooker, "What a book a Devil's Chaplain might write on the clumsy, wasteful, blundering & horribly cruel works of nature." Darwin could have easily illustrated his point drawing only from relations among family members: parents and offspring, brothers and sisters. Human reproduction fits Darwin's category of "clumsy" all too neatly. Both mothers and babies are ill suited for the ordeal, as a baby just a bit too large must squeeze through a birth canal just a bit too small. This requires that the skull of the fetus be sutured, allowing the bones to deform as they pass through the birth canal, and to make it work mothers must chemically soften the ligaments of their pelvis. And even with these accommodations, the baby still gets stuck some of the time. What a silly system.

There are prodigious examples of wasteful behavior. In grebes, pandas, and harpy eagles, parents throw away—deliberately abandon—perfectly good offspring. Parents can be blundering too. I could use myself as an example (babies really should come with owner's manuals) but instead will defer to brown-headed cowbirds, who escape the burden of parenting by laying their eggs in the nests of other birds. They do it crudely. They do not attempt to disguise their eggs' presence from the potential host parents, many of whom simply toss the alien eggs from their nest.

And relations between the closest of relatives can be unspeakably cruel. A baby black eagle if it is the second hatched in the nest—the beta chick—faces a life that is poor, nasty, brutish, and decidedly short. Its sibling, four to seven days older, greets it with repeated blows to the face and body. Each chirp or movement triggers another beating, whether the parents are present or not. At one nest, closely observed, the alpha chick delivered nearly three hundred pecks over beta's first 24 hours, another six hundred during the second day, and more than six hundred on beta's third and last day of life. By the end of day one, the victim was blood spattered about its face, beak, and eye; by day two its left eye was swollen shut; and by day three it lay weakly on the nest, awaiting death.

Such behavior certainly runs counter to our expectation that relations among family members, and particularly between parents and offspring, are nurturing, protective, and kind. This instead is a story of Darwinian live and let die. Parents and offspring are among the closest of relatives, and indeed their relationship can be warm and giving. Or it can be terribly selfish. Or anything in between. Over the last two decades evolutionary biologists have had to revise drastically their views about families. Families are not just simple, harmonious social units but are in fact far more interesting. Families serve as forums for rival evolutionary agendas where brothers and sisters, parents and offspring, cooperate, compete, deceive, and nurture. As family members we even compete with ourselves, in a war of genes derived from mother and father. There is strategy, counterstrategy, and layers of intrigue.

One of the most puzzling intrigues is this. Parents often make too many babies. This behavior is understandable in species such as ocean sunfish, which may produce 300 million eggs in a single clutch, or orchids, which make as many as a billion tiny seeds. Each egg or seed is a ticket in an evolutionary lottery, and the more tickets any individual holds, the better its chances of winning. Such organisms provide no parental care for developing offspring. Once the seed or egg is built and fertilized it is cast, with its parent's blessing, into the environment to find its own way.

When there is no parenting involved, producing as many progeny as possible maximizes the odds of success in the lottery. But parental overproduction is less understandable in birds, mammals, and some reptiles, as well as in amphibians, fish, invertebrates, and even plants that render considerable sums of parental care. Their offspring often depend on parents for long periods after fertilization: just think how long a human embryo/infant needs its parent(s) for its successful development and indeed its very survival. Yet even among organisms that provide such care, parents often produce more progeny, sometimes many more, than can ever survive to independence. Parents are optimistic, and this optimism leads inexorably to a secondary, downward adjustment of brood or clutch size by either infanticide or neglect. The obvious question is, why? Why do parent white pelicans and black eagles stand aside as older nestlings bludgeon younger brood mates? Why do hooded grebes, harpy eagles, or crested penguins abandon, bury, or eject viable eggs before hatching? Why do pandas and humans neglect newborns, and why do mice, owls, and burying beetles make routine meals of their own progeny? This nightmarish parenting not only occurs; it is widespread.

Parents also play favorites. They do so through birth or hatching delays, or by making some eggs or offspring larger than others, or even by fortifying some progeny and not others with steroid hormones. A first-laid black eagle egg hatches four to seven days ahead of its younger sibling, giving it a commanding advantage in the ensuing battle for life or death. Blackbirds and penguins make last-laid eggs bigger, sometimes much bigger, than first-laid eggs; litters of piglets and kittens often contain runts. Canaries and egrets dose some of their eggs with extra testosterone, making the chicks beg more and fight harder. Again, why?

One of the two central themes of this book is that nature provides useful models for the study of human behavior and reproduction. The following chapters provide a short tour of family relations in social animals with frequent detours along the way to explore where humans fit in. The tour's focus naturally reflects my own experience and bias. During the spring and summer months I work on birds. This has meant slogging through prairie wetlands in search of marsh-nesting blackbirds for more than a decade. Blackbirds comprise in many respects model systems. They are found everywhere in North America from the tree line south (and in the Caribbean and Central and South America too), are easily accessible, and occur in very large numbers. Unlike terns, their bills are not particularly sharp, or their aim particularly good. And they nest close to the ground, which is comforting for those of us who prefer to be close to the ground.

Birds, however, are far from the only kind of animal we encounter on this tour. During the long winter months in Winnipeg one has ample time to play hockey, root for global climate warming, and think longingly about blackbirds that nest in warmer places. Perhaps more important, one has the time to contemplate what avian models mean for other organisms with similar lifestyles, including us, and to write papers about this—sometimes with equations. The differences between humans and birds are obvious, the similarities less so, but the latter are what I find particularly compelling, and are what I explore here.

The second major theme of this book is that family harmony is not the default situation: that notion is what I call the family myth. Harmony is one possibility, but there are other, darker solutions too. Understanding families requires understanding that family dynamics are governed by the tension between conflict and cooperation among individuals who share common genes and experience.

I shall begin the tour by exploring the twin questions of how large a family should be, and how it should be structured. The answers require that we solve a twin paradox: why do parents produce too many babies (parental optimism), and why do they play favorites (parental favoritism)? The resolution of these paradoxes is key to understanding family relations, as parental optimism and favoritism are the direct antecedents of much of family conflict.

Chapter 2

The Optimistic Parent

The Evolution of Family Size

> *"Before I draw nearer to that stone to which you point," said Scrooge, "answer me one question. Are these the shadows of the things that Will be, or are they shadows of the things that May be, only?"*
>
> —Charles Dickens

In *A Christmas Carol*, Dickens's curmudgeon confronts the question of whether the future is already foretold. Scrooge himself puts it best: " 'Men's courses will foreshadow certain ends, to which, if persevered in, they must lead,' said Scrooge. 'But if the course be departed from, the ends will change. Say it is thus with what you show me!' "

Much would change if we indeed knew what the future would bring. Why carry a spare if tires are puncture proof? Why buy fire insurance for a house built of asbestos? Health insurance perhaps, fire insurance no. We use backups as a hedge against the uncertainty and prudently plan for possible failure. In a perfect world, parents would face no uncertainty about the ideal size and composition of their family. Forthcoming resources would be wholly predictable; the sex of their progeny would be known in advance; and the health and survival of all offspring would be assured. But in an imperfect world, parents face the challenge of rearing a family in the face of ecological and developmental uncertainties. The prudent parent establishes contingencies for possible failure.

Parental optimism—creating more incipient offspring than are likely to be sustained to independence—is a multipurpose solution to many of the vicissitudes of family planning. The general problem is simple. Parents do not have sufficient information about the future. Like Scrooge, parents might make different decisions if they knew to which ends a decision would lead. If a parent bird or mammal knew that food supplies would be lower than average, it might start with a smaller clutch or brood. But with incomplete information, parents must hedge their bets.

Many of the key unpredictabilities of family life begin from within. Some offspring are born with congenital defects. A sperm or an egg may carry a chromosome too many, or a developing embryo may be injured by a toxic chemical, inducing a mutation. Other threats to family harmony are external. Some offspring will be afflicted by injury, parasites, or pathogens during the course of development. Others may prove unaffordable within the limits of their parents' current budget. Famine may strike unexpectedly, or cold weather, elevating the costs of keeping the children warm. And key sources of environmental variability extend beyond the period of parental care. What sort of environment will the offspring encounter at independence? Large, fat offspring may do better if food conditions are stringent, but if food is plentiful, the offspring might get by with less. Or even worse, being large and fat may also mean slow and clumsy, and vulnerable to predators. These are uncertainties of ecology and development with which all parents cope.

Both the size and structure of families are shaped by such ecological and developmental uncertainty, and a contingent view of families helps to make sense of otherwise puzzling and seemingly wasteful features of reproduction. Brown boobies lay two eggs but raise only one chick. Sand tiger sharks murder and eat brothers and sisters in utero. Embryos of pronghorn antelope impale sibs in a narrow uterus with a spear of dead tissue. Nestling bee-eaters use modified bills, and piglets modified teeth, in fatal sibling rivalries. Parents of beetles, bees, ants, owls, and mice feast on their own children.

A common thread for this grim pattern is that parents make too many offspring. This is the result of parental optimism. Parents routinely create more incipient progeny than they can—or will—rear to independence. In a perfect world, parents would create exactly the right number of offspring. All would be born healthy, without intrinsic defect, and all would receive ample provisions and lead happy lives. If only it were so. Instead, nature reveals a Hobbesian world, where the lives of many larvae, embryos, infants, children are nasty, brutish, and short. No more extreme example can be found than in obligate brood-reducing species, for which the early death of one's progeny is a mandatory component of reproduction.

The Puzzle of Obligate Brood Reduction

Brood reduction is the loss of one or more dependent offspring due to filial infanticide, parental neglect, or fatal sibling rivalry. It is a remarkably widespread though not always mandatory habit, occurring at some time in nearly every species that rears multiple, simultaneous offspring. *Obligate* brood reduction is an extreme form of the practice, in which at least one offspring is doomed to perish during the period of parental care.

Harpy eagles are giant raptors that patrol the forests of South and Central America. They lay two eggs in treetop nests, and when the first hatches, the second egg is buried. Hooded grebes build floating nests in freshwater marshes of South America, and they too lay two eggs. When the first hatches, the family swims off, abandoning the unhatched chick. Pandas routinely give birth to twins but nurse only one. The second is dropped to the ground and left to die. Parents in these species deem the "surplus" progeny expendable.

In a variety of predatory birds—pelicans, eagles, boobies, cranes—siblings play the role of executioner. Two eggs are laid, and they hatch at unusually long intervals for birds, several days apart. The first chick to hatch gains an immediate size and strength advantage over its younger sibling. When the second chick hatches, it faces an unrelenting assault from its brother or sister that ends only when one, almost always the younger, dies. The paradox of obligate brood reduction is this. If one offspring is sure to die, why produce two? Why do white pelicans and black eagles lay the second egg? Natural selection abhors waste, and this pattern of reproduction is inherently wasteful. Obligate brood reduction is one extreme of a strategy of parental optimism, and as is so often true, the extreme proves instructive.

The paradox of obligate brood reduction is easily resolved. Like Scrooge, parents are not naturally inclined to be spendthrift, nor do they know what the future holds. But animal parents do not benefit from the guidance of clairvoyant apparitions and must hedge their bets accordingly. Breeding eagles, pelicans, pandas, boobies, cranes, and even pine trees plan for failure—any failure of their progeny to develop normally and thrive after hatching, birth, or germination. They do so by purchasing a cheap insurance policy: a spare embryo. Should the first egg fail to hatch, or if the firstborn/germinated suffers an innate defect, parents have a backup ready to insert in its place. But if all is well with the first, the backup becomes expendable. This is a story of Darwinian live and let die. Parents and offspring are among the closest of relatives, but even this does not protect later-borns from the harsh calculus of family economics. Otiose offspring are abandoned or killed.

How Many Babies?

> *The cost of a thing is the amount of what I will call life which is required to be exchanged for it.*
> —Henry David Thoreau

The paradox of obligate brood reduction raises a further question. If all are born healthy, why not capitalize on the good luck and raise the extra progeny? This addresses a central question of life-history theory: what sets the optimal family size? The answer revolves around two key tradeoffs. The first is between the number and quality of offspring, the second between current and future reproduction. All parents work within a budget; they are constrained by what economists refer to as the principle of allocation. A dollar or kilojoule spent on one item cannot be spent on another. Thus there is a necessary tradeoff between offspring size and number. Progeny can be big and few, small and many, or something in between. Humans favor a few large and very expensive offspring. Each child that survives infancy receives enormous quantities of parental care and stands an excellent chance of surviving to adulthood. Mice, ducks, and beetles favor larger broods of not-so-expensive offspring. Each progeny receives less parental care and is less likely to survive to adulthood as a direct result. Since a larger brood means a lower per capita investment in each offspring, more is not always better. Indeed it is often much, much worse.

Here it is instructive to look at giant pandas. A 90-kg mother panda is nearly a thousand times larger than her newborn infants, which average 100–150 g. But the imbalance does not last: by six months of age, the infant has increased more than a hundredfold in mass, and it will not be weaned for yet another three months. Pandas live and breed at a leisurely pace, in large part because their diet—bamboo—is poor in nutrients and energy. Rearing twins, therefore is a significant though not impossible challenge. Mother pandas can and occasionally do raise twins in the wild, but not without cost. Twins of captive pandas grow considerably slower than singletons, unless their diet is artificially supplemented. Although at birth the extra infant is easily sustained, the prospect of pending competition between the infants looms in the not-too-distant future (see chapter 3 for discussion of pending competition). More is not necessarily merrier for the mother panda, and very likely two relatively malnourished twins are no better than a single robust infant. Pandas are in it for the long haul, as a female panda may raise half a dozen infants over her twenty-odd-year life span. This brood size and lifetime reproductive success resembles that of many other large and long-lived mammals, including our closest living relatives, chimpanzees, and even humans living in hunter-gatherer societies.

For humans too, more is not merrier. Between 95 and 99 of every 100 live births result in just a single baby. We can be thankful that the modal brood size for humans is one. Even with the miracles of modern science and technology, twins are more than a handful. Intriguingly there are many more multiple conceptions than multiple births, a statistic that hints at a more sinister story (a topic examined in chapter 8). Similarly albatross, kiwis, and California condors lay but a single egg, and elephant seals, horses, blue whales, and black rhinos give birth to single offspring. Smaller birds and mammals are more ambitious with brood size. House cats have litters of two to six and dogs four to ten, depending on breed. Mice and opossums have up to a dozen babies at a time, rats as many as fourteen, and pigs up to sixteen. The same holds true for birds. Pigeons and doves are content with two eggs, sandpipers four, and blackbirds three to seven. In species that feed their own nestlings, the dozen or so eggs of chickadees and titmice is the upper limit. For ducks, rails, and partridge, which rear precocial offspring, up to thirty eggs may be laid, although it is not always possible to tell if they come from one female.

The ornithologist David Lack identified the key tradeoff between offspring size and number. Parents can make many small offspring, each with low prospects of survival, or a single large offspring with excellent prospects for survival, or something in between. Number is traded off against quality. The object of the evolutionary game is not to maximize offspring numbers but to maximize the number of *surviving* offspring, those individuals most likely to continue the evolutionary legacy. Larger broods are disfavored if such a strategy leaves malnourished progeny with poor prospects for survival. The resources that parents have to invest in their offspring are finite (a fixed pie), and the evolutionary problem that parents must solve is how to divide the pie among their current brood: should there be three slices or four, and should they all be the same size? The size of the pie to be divided is set in part by the second key tradeoff, that between current and future reproduction. More progeny now generally mean fewer later.

Organisms that invest massive amounts of parental care in each offspring necessarily raise small broods. California condors produce one chick at two-year intervals, a frequency similar to that of some large whales. These parents mete out parental investment carefully and may live for a century or, in the case of bowhead whales, perhaps two centuries. Conversely, great tits may rear a dozen chicks in a single clutch and sometimes produce two clutches in a single summer. But half of these parents will die before the next breeding season arrives. House mice produce litters of up to a dozen pups at monthly intervals but live on average for a hundred days. Small

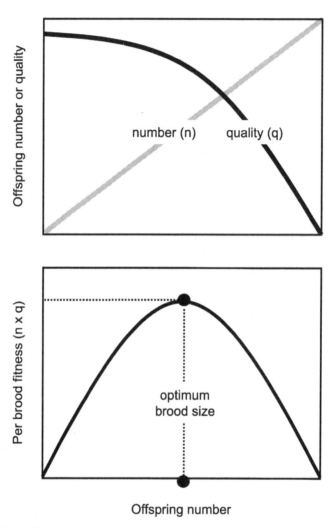

Figure 2.1. The Lack brood size
The British ornithologist David Lack argued that producing more offspring is not necessarily better, because as brood size (*n*) increases (*gray line in top panel*), the quality of offspring (*q*) declines (*black line in top panel*). The product of quality and number (*black line in bottom panel*) is what determines the fitness of the brood to parents. The optimum brood size is that where this product reaches a maximum, indicated by the black dot.

birds and mammals live fast and die young, large birds and mammals quite the opposite.

Although they produce too many offspring, parents are careful not to waste precious resources on them. Birds and mammals are generally conservative in providing for their broods, with larger species more so, and usually do not work at or anywhere near the maximum levels possible when raising a family. This habit can produce the appearance of indolence, and parents can indeed be lazy, sometimes impressively so. But they are lazy for good reason. We see this most clearly when brood size is experimentally enlarged. At least over the short term, parents are willing and able to step up work levels to raise more babies. Such spare work capacity is useful to cover temporary and unpredictable periods of food shortfall or elevated demands from offspring.

Ospreys, which are large fish-eating hawks that erect nests in tall trees overlooking waterways, provide a conspicuous example of parental sloth. A pair works in tandem to rear one to three chicks over a three-month nestling period. Males and females divide the labor, with dad doing most of the hunting for the family and mom performing most of the incubation of eggs and brooding of chicks. Neither parent appears particularly busy, and male ospreys make even Homer Simpson look ambitious. Every two or three hours, the male leaves his perch beside the nest, hunts for a short time, and then returns with a fish. The rest of the time he sits near the nest, sleeping, yawning, scratching, and staring vacantly at the clouds. And what does he do when one of the nestlings is malnourished, becoming a target for bullying by its older, larger, and stronger siblings? Mostly the same. Most ospreys lay three eggs, but only one in ten nests fledges three chicks. At most nests the last-hatched chicks fall victim to a combination of starvation and sibling aggression. Fathers, who have ample time to step up their hunting effort, do nothing to prevent the death of their offspring.

When one examines the array of organisms that spend substantial time and effort rearing their offspring—certain insects, fishes, and amphibians, and nearly all birds and mammals—this pattern is surprisingly common. Parents create more eggs or neonates or larvae than they are willing to rear and stand by as the surplus perish. Experiments show that most of the time parents are quite capable of rearing more progeny than they do. Gannets are large seabirds that lay a single egg. When ornithologists experimentally provide a second egg, most gannets manage to rear two chicks to independence. Why then do parents not work harder? The answer is disarmingly simple. Sloth pays.

The regulation of family size is a central question for evolutionary ecologists, and birds—because they do not hide their progeny in inaccessible wombs or dens, or out of sight underwater—are favorite subjects for study. Ornithologists have asked many times whether parents can raise more nestlings than they normally do, and almost always the answer is the same. Yes! So the pattern seems well established: greedy parents initiate larger families than they rear, and then children in peril are neglected or worse. Parents also add malice to complete a Machiavellian trinity of unholy parenting: greed, sloth, and now favoritism.

Avian Families

Songbirds lay anywhere from two to twelve eggs that hatch after ten to thirty days of incubation. The nestlings enter this world naked and helpless and require constant brooding and feeding. One or both parents may provide direct care for the offspring, and if only one, it is usually the mother. Even though parents have an equal genetic stake in all their progeny, not all offspring are created equal. In true Orwellian fashion, some are rendered more equal than others by parental fiat.

Altricial birds (those that raise helpless young) usually hatch their offspring asynchronously by starting incubation before the clutch of eggs is complete. Birds can lay at most one egg per day. If incubation commences with the final egg, all nestlings will hatch in rough synchrony, usually within a few hours of each other. But if incubation commences with the penultimate egg, as is often the case, all but the last will hatch together. That last laid egg will hatch a day later. And a one-day head start can quite literally make the difference between life and death. In red-winged blackbirds, the mortality rate for first-hatched or "core" offspring prior to leaving the nest is low: between one in ten and one in twenty-five perish prematurely. For last-hatched or "marginal" offspring that figure rises sharply, to between one in two and one in three nestlings.

Core and Marginal Offspring

I shall use the terms *core* and *marginal* offspring throughout this book. The terms recognize that some offspring, the core brood, are more important to parental interests than others—the marginal brood. Some of the time the core and marginal offspring look the same, and the distinction is more or less arbitrary. Such is the case in burying beetles that lay eggs on a mouse carcass. Some larvae, the

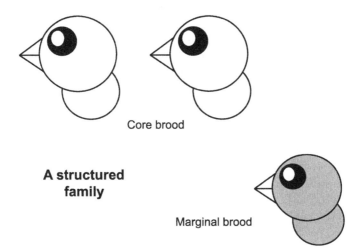

A structured family

Core brood

Marginal brood

Figure 2.2. The structured family
Parents often structure families into core and marginal offspring that serve different functional roles.

core brood, will complete development normally, but for others, the marginal brood, a very different fate is in store. Their parents will eat them. At the time of egg laying it is not clear which eggs will give rise to core, and which to marginal, progeny.

In organisms such as birds, parents impose a physical handicap on marginal progeny, and the marginals are easily recognized. Marginal nestlings of red-wing blackbirds, for example, lag behind their larger core siblings in both size and development through no fault of their own, and the manipulative parents may even compound their woes by giving a chemical edge to the privileged core brood: an extra dose of testosterone that renders chicks more belligerent at feeding time. Conversely parents may narrow the competitive edge by giving the steroids preferentially to the marginal chicks, or by making last-laid eggs, and hence hatchlings, larger. Red-winged blackbirds do both.

The obvious differences notwithstanding, there are intriguing parallels between humans and altricial birds. Human offspring are born naked and helpless, and require constant care when young. Births normally occur at intervals, albeit measured in years instead of days, and multiple births occur relatively rarely. Between one in twenty and one in one hundred live births result in twins, depending on population—women of African descent are more prone to twinning, East Asian women less so. The rate of twinning also depends on maternal age, height, and weight. Older, taller, and fatter women are more likely to give birth to twins.

Synchronous broods in birds, and closely spaced or multiple births in humans, elevate the demands that offspring impose on parents. We can observe the results more easily in birds. When ecologists experimentally synchronize naturally asynchronous broods, the usual result is that synchronous broods produce more but lower-quality offspring than asynchronous broods. Here parent birds are squeezed by the principle of allocation: one can make more or better offspring but not both. In asynchronous broods, last-hatched nestlings consume less food than their core counterparts, particularly when food is short. The result is a smaller number of well-nourished core sibs and one malnourished or dead marginal offspring. In experimentally synchronized broods, parents in some cases assume the extra costs by working harder to rear all their nestlings, but do so at a cost to themselves and their future offspring.

Asymmetric Sibling Rivalry Creates Disposable Offspring

Is a parent's love a genetic imperative? A biological parent is equally related to all his or her offspring. So why play favorites? The answer is that doing so allows parents a wider range of options. Parents can add one or more extra offspring to their brood and trim the excess quickly if necessary, creating in effect a caste of cheap, disposable offspring. Again, the dynamics are easiest to observe in birds.

Marsh-nesting blackbirds are a model system for the study of parental favoritism. They live in capricious environments that vary hour to hour, and parents cannot predict the future more than a few days in advance. A mother blackbird usually starts incubation of the clutch (she receives no help for this task from dad) before she has finished egg laying, and thus her eggs hatch asynchronously. The marginal blackbird nestlings that result receive smaller food shares than older, larger core progeny, particularly when the brood size is large. This mechanism establishes an important competitive asymmetry between the core and marginal brood.

The presence or absence of a marginal nestling in blackbirds has almost no detectable effect on the growth or survival of core offspring, but the reverse is not true. The fate of marginal offspring hinges on the fate of core offspring. Should a core nestling fail early, or be removed experimentally, the growth and survival rates of marginal nestlings rise sharply. This same pattern occurs in an even more extreme form in obligate brood-reducing species such as black eagles, white pelicans, and brown boobies. These birds lay two eggs but at most one nestling survives the crucible of sibling competition. Shortly after hatching, the nestling from the last-laid egg is bludgeoned quickly and ruthlessly to an early death by its older, stronger sibling. It survives only if the first-laid egg fails to hatch a viable nestling. Parent birds in these species are quite capable of rearing more than one chick but choose not to. Maximizing the number of current offspring is not the key to success in these large, long-lived species. Rather, they are running a reproductive marathon, and are guarding future reproduction by rearing only a single chick to independence in each breeding attempt.

The Evolution of Family Structure

How many babies should parents rear? The evolution of family size is a fundamental question of life history. But how do we measure family size? Avian ecologists have long used such logical and obvious measures as the numbers of eggs, nestlings, or fledglings. But by imposing handicaps on certain of their progeny and not others, bird parents establish competitive asymmetries among the offspring. Not all offspring are created equal, so perhaps they should not be counted as equals when measuring family size. The phenotypic (physical) differences created due to hatching asynchrony and differences in egg or neonate size, or in hormonal boosts, create a structured sibship, effectively dividing the brood into the have (core) and have-not (marginal) offspring. The corollary of this brood structure is that core and marginal offspring serve different functions.

The core brood represents the minimum subset that parents can expect to rear under the normal range of ecological conditions encountered. Often, though not always, core brood members are protected from competition within the brood via a handicap imposed on the marginal progeny. When so protected, members of the core brood enjoy secure prospects for growth, survival, and ultimately fitness even in the face of variable environmental and developmental conditions. The handicap serves to buffer the core brood from both extrinsic (e.g., weather, food levels) and intrinsic sources of variation (e.g., failure to hatch, congenital defect). What happens to the core brood is largely, albeit not completely, divorced from what happens to the marginal brood. The opposite, however, is not true.

The fate of marginal progeny rests largely on the fate of the core brood. Obligate brood-reducing species laying two eggs provide the most obvious example. Should both hatch, the first-hatched chick usually murders its younger sibling. The parents provide an overwhelming advantage to the first-hatched chick by hatching their two eggs at long intervals, routinely three to seven days apart. The ensuing battle between nestlings is bloody, one-sided, and mercifully brief. Here the fate of the marginal chick rests almost solely on the fate of its core sibling, whereas it matters little to the core chick whether the marginal chick hatches or not.

The same asymmetric sibling rivalry holds even in pacifist blackbirds. They practice facultative brood reduction, that is, brood reduction that occurs sometimes but not always. As it is with unions when workers must be laid off (last in first out), so it is with blackbirds. The last hatched are almost always the first to die. But unlike the doomed marginal offspring of obligate brood-reducing species, the marginal blackbirds sometimes survive alongside their older, stronger core brood mates.

This conceptual shift, from the traditional assumption that all offspring are equally valuable to a view of some progeny being more valuable than others, has important practical implications. It is often far more useful to refer to the size of the core and marginal brood separately than to talk about a unitary measure of brood size: a brood of two core and two marginal offspring can be very different from a brood of four core offspring, even though the overall brood size is the same. Family structure can be just as important as nominal family size.

The core-marginal dichotomy, though it can be broadly applied to animals as diverse as sharks, pandas, grebes, and humans, emerged from empirical studies of altricial birds. Among many species of songbirds, parents hatch their nestlings semisynchronously—that is, in a clutch of n eggs, n −1 hatch more or less simultaneously (the core brood); and the last-laid egg hatches one or more days later (the marginal chick). In other cases such as hawks, owls, and parrots, most or all of the brood may hatch asynchronously, creating a complete brood hierarchy. Here the distinction between the core and marginal brood blurs. The first-hatched chick or chicks remain the core brood, and later-hatched chicks in a technical sense would be defined as the marginal brood, but the most senior marginal chick may closely resemble the core chick in its trajectory of growth and survival, the last-hatched marginal chick not at all. Obviously, in cases where there are multiple marginal offspring the distinction between core and marginal glosses over sometimes important differences among marginal chicks, but the dichotomy is still a substantial improvement over the traditional view of treating all progeny as more or less equal in the eyes of their parents.

The core-marginal dichotomy finds an immediate human parallel in the phenomenon of primogeniture, according to which the eldest son is designated as the heir to the family estate. Simply by an accident of birth one son is preferred over another (and over any daughters) under such systems. The family has a vested interest in ensuring the success of the firstborn male, and later-borns are forced into different roles (heir to spare) and occupations. Again, not all offspring are created equal.

The core-marginal dichotomy can be brought into service to resolve otherwise paradoxical arrangements within families. Crested penguins have quite deservedly earned the adjective "bizarre." They lay two eggs that are more different in size than those of any other bird: a small, first-laid egg that is only 60% of the size of the larger, second-laid egg. These are the A (first) and B (second) eggs. But the second-laid B egg hatches first, and the first-laid A egg hatches second. Among nearly all other birds, eggs hatch in more or less the sequence they were laid. By custom we call the first-hatched the A chick and the second-hatched the B chick; thus the crested penguin's B egg produces the A chick, and the A egg the B chick. And in most cases it is the B egg/A chick that survives, outcompeting the A egg/B chick. The core-marginal dichotomy tidily resolves the problem. The second-laid (large) egg is the core offspring, the first-laid (small) egg the marginal offspring.

What Is Parental Optimism?

Parents produce more offspring, often many more, than will be reared to independence in organisms as diverse as sharks, beetles, and birds. The resulting mismatch between food supplies and resource demands leads directly to sibling rivalry with often fatal consequences. Snails, frogs, and fish cannibalize sibs. Pronghorns impale brothers and sisters in utero, and blackbird nestlings wither and die in the face of stiff competition for food and warmth from older, stronger siblings.

Yellow-headed blackbirds provide a tidy illustration of parental optimism at work. These handsome birds breed in prairie wetlands across central North America. They weave nests of reeds, in which they lay clutches of usually three to five eggs that are hatched asynchronously. The first two eggs normally hatch together, followed by the others at one-day intervals. As a direct consequence the first-hatched, core nestlings gain a one- to three-day head start on their later-hatched nest mates, an advantage that spells the difference between a cosseted existence with excellent prospects for growth and survival (core sibs) and a difficult struggle for survival. When a bout of cold, wet weather arrives, core sibs obtain warm spots at the bottom of the brood huddle and secure first access to food if and when it does arrive. Last-hatched chicks perish first when ecological conditions deteriorate.

Yellowheads are also sexually dimorphic (males and females have distinct forms). By the time the chicks are ready to leave the nest, at about twelve days of age, males are half again as big as females. The big bodies of male nestlings are expensive to maintain: two males are roughly as costly as three females. When food is short, being male becomes a problem, and here the role of hatching asynchrony is key. In bad years, parents cannot raise more than a single male, though they can raise two females. Yellowheads have evolved a neat trick to deal with this contingency. They appear to avoid placing more than a single male in the core brood. Any additional males are placed in the marginal brood, where they are easily eliminated when food is short. In an experiment that eliminated this natural hatching asynchrony during a particularly bad food year, being a male in a synchronized brood proved a lethal combination. No males survived in any synchronous broods, whereas in naturally asynchronous broods some, though not many, survived. Brood synchrony reduces the efficiency of brood reduction, and when resources are scarce, multiple males become unaffordable. Delaying the onset of brood reduction—a cost of synchrony—rendered maleness an unaffordable luxury, and all males perished.

But during warm, dry years when brood-rearing conditions are favorable, marginal chicks thrive—even males—and are reared alongside the core brood. And in all years the prospects for marginal chicks improve markedly if a core chick fails to hatch or perishes early. Here the marginal chicks serve multiple functions for parents. They replace failed core progeny, allow for expensive males to be trimmed from the brood under conditions of stringency, and providing a reproductive bonus when brood-rearing conditions are favorable.

Yellowhead families are typical of those that span the taxonomic spectrum, from sharks and frogs to pigs and pandas to eagles and pelicans. Parents benefit by creating too many incipient offspring and playing favorites. As a direct consequence of this initial parental optimism and subsequent favoritism, legions of incipient progeny march forth to be decimated by parental neglect or malice. Why are so many organisms so spendthrift with their progeny? The evolutionary biologists Stephen Stearns and Jan Kozlowski provide the answer to this question. Eggs and embryos are cheap. It is the subsequent parental care that is costly. Thus parents routinely start with more progeny than will ever survive to independence, and they deploy this surplus to serve multiple adaptive functions that I shall explore below. But in organisms such as birds and mammals with expensive parental care, offspring do not remain cheap for long, and thus mechanisms to trim brood size are a necessary component of a strategy of offspring overproduction.

Why Parental Optimism?

The routine overproduction of progeny by parents is a strategy my colleague Douglas Mock and I refer to as parental optimism. Paradoxically, under such a strategy some offspring best serve parental interests by dying. This view offends the logic that more is better. But as David Lack long ago suggested, there is more to being a parent than simply rearing as many offspring as possible. Rather, it is about rearing the offspring that are most likely to transmit the parents' evolutionary legacy. And initial parental optimism creates manifold ways to do this.

The optimistic parent is manipulative, and core and marginal offspring are deployed to fill different roles. Some are destined to live independent lives as future reproductives. These are the favored core brood. Other offspring, though, serve supporting roles. They may guard the core brood, or serve as replacements or upgrades, or even be served as food. These are the marginal brood.

The core offspring represent the minimum subset of progeny that parents desire to rear in any breeding attempt. The marginal offspring are the supernumeraries that augment the core brood. It is the overproduction of marginal progeny that at first glance appears puzzling, a puzzle that is resolved by considering a trio of incentives for parents to overproduce initially and then later trim family size. Marginal offspring give parents more options for (1) tracking resources that vary unpredictably, (2) replacing failed or feeble core progeny, and (3) facilitating other family members.

Tracking Erratic Resources

> *He chose to be rich by making his wants few, and supplying them himself.*
>
> —Ralph Waldo Emerson on Thoreau

The balance between resource supplies and demands is often precarious. It rests not only on what resources (chiefly food) are available to parents but also on how much their offspring need. Both sides of

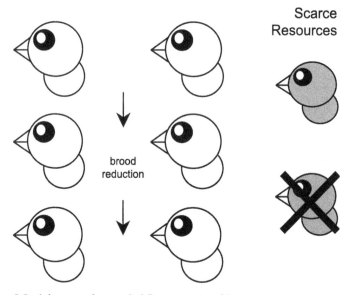

Figure 2.3. *(above and opposite)* **Resource tracking**
Under resource tracking, the marginal offspring are raised in addition to the core brood when resources are plentiful, but become expendable when resources are scarce.

this simple equation can vary unpredictably. Parents would ideally like to know if good or bad conditions lie ahead, though such information is often unavailable. Consider the following example.

Bald eagles in northern Saskatchewan raise eaglets on the fish they catch in local lakes. But when they lay their eggs, those lakes are covered by snow and ice concealing the food supplies that lie beneath. Eagle parents have an information problem. They do not know what the spring thaw will bring, and in the absence of perfect information about forthcoming conditions it pays to hedge. Environmental conditions might be unusually favorable and extra offspring of value. Or they might be unusually bad, favoring a small family size. Parents can keep their options open by producing one or more extra and potentially surplus progeny. If beneath that ice and snow lies a bumper crop of suckers, walleyes, and pike, then parents will have an easy time rearing a trio of eaglets, and the marginal offspring can be cashed in for a reproductive windfall. But if fish populations have been decimated by a severe winter, the eagle brood can easily be trimmed to two or even one to restore the balance between fish supply and brood demand.

This balancing act is resource tracking. Its adaptive value holds not only for birds but also for plants, piglets, polar bears, and even

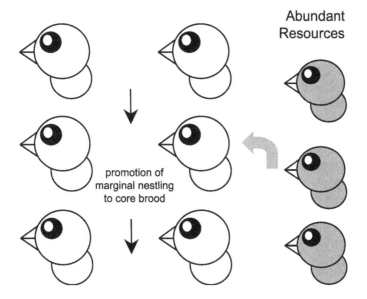

Abundant Resources

promotion of marginal nestling to core brood

humans. Parents add marginal offspring to the clutch or brood just in case they are needed, and trim the surplus as necessary. The diagnostic feature of resource tracking is that the fate of marginal offspring improves as that of core offspring improves.

The strategy, in general, is that just described for bald eagles. Parents create more progeny than will likely be reared. If resource supplies meet or exceed brood demands, the surplus marginal offspring are cashed in—kept alive for their extra reproductive value. But if food supplies fall short, offspring numbers require trimming. Ornithologists refer to this process as brood reduction, defined as offspring mortality that occurs as the outcome of sibling competition or, less frequently, filial infanticide.

The study of brood reduction in birds has a long tradition, dating back to the first half of the twentieth century, when the British ornithologist David Lack suggested that parents may create an optimistic clutch size (though he did not call it that) and trim brood size downward if food supplies are insufficient to meet the demands of the family. Lack thus framed what others termed the resource-tracking hypothesis, one of the trio of adaptive explanations for the production of surplus offspring within the framework of parental optimism. And according to Lack, hatching asynchrony assists the process by producing easily disposable runts that can readily be eliminated if and when food is short.

Lack's logic was and is sound, though too simple. Much subsequent work by ornithologists has shown that hatching asynchrony lies amid a network of coadapted (mutually evolved) traits, as the Australian biologist Robert Magrath has correctly described. The extent of hatching asynchrony is affected by clutch size, parental age, workload of the parents, rates of predation and of whole-brood loss, the sex ratio of the brood, nest size and the size of the parent, the availability of nest sites, polygamy, prevailing food levels, rates of egg failure, and sibling competition. Not surprisingly, when avian ecologists set out to test Lack's hypothesis as the sole and sufficient explanation for avian hatching asynchrony, they failed. As H. L. Mencken once noted: "For every complex problem there is a solution which is straightforward, simple and wrong." This pithy remark seems apt for most ecological phenomena, and particularly for the study of avian hatching asynchrony.

Field tests of Lack's hypothesis—conducted chiefly by experimentally reducing or eliminating hatching asynchrony by swapping eggs or chicks across nests—yield uneven results. Broods that are experimentally synchronized tend to produce more chicks that survive to leave the nest, but those chicks are malnourished relative to chicks in normally asynchronous broods and probably do not survive as well afterward. In contrast, the natural pattern of hatching asynchrony yields fewer but higher-quality offspring. Critics of Lack's hypothesis have argued that hatching asynchrony should result in more, not fewer, surviving nestlings, though why this should be so is unclear to me. If hatching asynchrony is indeed designed to facilitate brood reduction, asynchronous broods should produce either the same number or fewer nestlings than otherwise equivalent synchronous broods.

There is one caveat. Hatching asynchrony may avert the loss of the whole brood to starvation, which might occur in a brood of evenly matched competitors. This was indeed part of Lack's original argument, though he probably overstated its import. Whole-brood loss due to starvation is relatively uncommon in birds. Among blackbirds, for example, it occurs only rarely. The symptoms are easily recognized. Entire broods grow slowly, and one by one the individual chicks weaken and die.

Norwegian biologists Tore Slagsvold and Trond Amundsen have updated Lack's original argument and have turned the focus from the risk of whole-brood loss to the effect of hatching asynchrony on offspring quality. They argue that hatching asynchrony in birds serves to produce fewer but higher-quality progeny than would be produced in an equivalent synchronous brood. The early brood reduction afforded by asynchrony diverts more food to the survivors, the logic being that having fewer but robust offspring is preferable to having more but malnourished offspring. Though more progeny may leave the nest from a synchronous brood, fewer survive to breed or, if they do survive, they may be less attractive to members of the opposite sex. This "quality assurance" hypothesis thus places Lack's argument in a life-history context. The process does not end when the chicks leave the nest but rather is just beginning.

Brood synchrony sometimes imposes a further cost on parents, as exaggerated sibling competition—a consequence of evenly matched competitors—can elevate the food demands of the brood. Experimentally synchronized broods of cattle egrets fight more and eat more than normally asynchronous broods, forcing parents to work harder to raise the same number of chicks. In such cases the effects on the parents' future reproduction must be included in the evolutionary accounting. Superimpose on this another important factor, the variation in environmental conditions within and across breeding seasons, and the costs and benefits of hatching asynchrony to parent birds become very difficult to measure. But no one ever said science was easy.

Resource tracking is not the only reason to produce surplus marginal progeny. Nature is ever resourceful, and marginal offspring, once created, can serve more than one function simultaneously. One additional function of surplus marginal progeny is to stand in as potential replacements for the core brood.

Replacement

The second prong of the trident of parental optimism is replacement. The production of marginal progeny also allows parents to hedge against the uncertain development of their brood. Some offspring perish unexpectedly and prematurely, while others are born with congenital defects. And still others are stricken by parasites or pathogens early in life. The presence of one or more marginal progeny allows parents the option of *replacing* these failed or feeble brood members without delay.

There are two adaptive advantages to replacement. The first is insurance: the reserve offspring can be used as a backup against the early failure of a core offspring. The second is progeny choice: low-quality core offspring can be replaced by higher-quality marginal offspring. Here parents can use the crucible of sibling competition to identify offspring with the best fitness prospects, weed out inferior progeny, and direct their investment toward superior progeny. Inferior may be defined as a genetic absolute (e.g., congenital defects) or may be relative to current environmental conditions (e.g., males

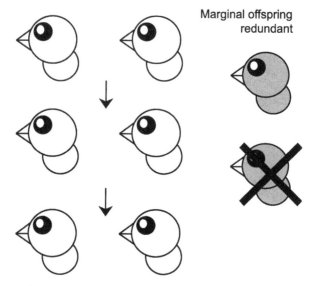

Figure 2.4. *(above and opposite)* **Replacement**
Under replacement, a marginal offspring substitutes for a failed or low-quality core offspring. If all core offspring are healthy, the marginal offspring is redundant and may be eliminated by brood reduction. Otherwise its continued existence may impose a crowding cost on the core brood.

might be favored if the mother is healthy or if food is abundant). Parents nurture the strong and discard the weak.

The diagnostic feature of replacement is that the fate of marginal offspring is contingent on that of the core progeny. If all core progeny are healthy and strong, the replacement value of marginal progeny is nil, and if replacement is the sole value of marginal progeny, they are no longer needed. Thus in obligate brood-reducing species, mechanisms to eliminate the surplus exist. If, however, marginal offspring serve multiple functions simultaneously, their continued presence may still serve parental interests. One of these parental interests is to nurture other family members.

Facilitation

The third incentive for producing extra progeny is facilitation. In this case marginal offspring assist other family members, often by dying. The extras may assist in group defense of the brood, as in social insects. Or they may help to conserve precious heat by huddling, as in litters of mammals and broods of birds. Or they may

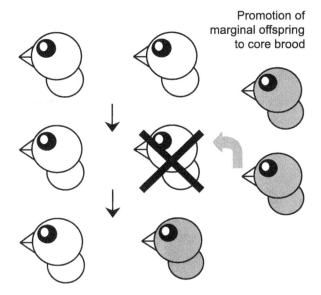

Promotion of marginal offspring to core brood

help to overcome defensive barriers as in parasitoid wasps, whose extra larvae help to overcome the host's immune system, though only one will ultimately survive. Or they may serve as soldiers or police. In other parasitoid wasps, a small proportion of the brood is transformed into warriors that combat potential predators or competitors. Some of these warriors also police the brood to trim surplus males. Or, most spectacularly, the marginal progeny may serve as food. In bees, ants, sharks, frogs, and occasionally birds and mammals, the surplus offspring are eaten by parents or siblings, providing a nutritional boost to the survivors. Even in death, marginal progeny yield benefits to the remaining family members.

Multiple Incentives for Parental Optimism

In some circumstances just a single function for marginal offspring dominates. This is the case in obligate brood reducers such as white pelicans: the primary value of their marginal offspring lies in the replacement of failed or feeble core progeny. But the trio of functions for marginal offspring under the umbrella of parental optimism—resource tracking, replacement, and facilitation—are not mutually exclusive. Indeed, a single marginal offspring can serve all three functions simultaneously.

Among marsh-nesting blackbirds the marginal offspring allow parents to track varying and unpredictable ecological conditions

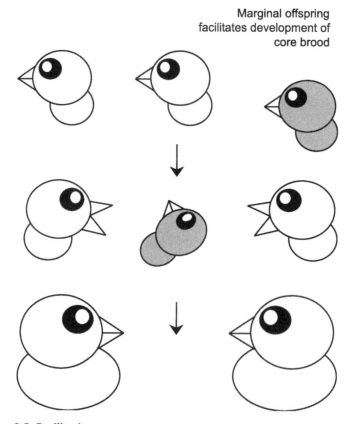

Marginal offspring facilitates development of core brood

Figure 2.5. Facilitation
Under facilitation, the presence of the marginal offspring enhances the fitness of other family members, and in particular members of the core brood—e.g., by serving as food.

with their brood size, and thus yield extra reproductive value. The marginals also serve an insurance function, replacing core eggs that fail to hatch, or chicks that perish early. They also serve a progeny-choice function, allowing for expensive male chicks to be jettisoned when brood-rearing conditions are poor. And when the hatchlings are young and the weather cool, the marginal chicks serve as living blankets for core chicks that huddle underneath, thus providing facilitation value. Among American kestrels, marginal offspring yield a reproductive bonus when small mammals are abundant, serve as a replacement for core eggs that fail to hatch, and are served as food (facilitation) when small mammals are scarce.

Plants also reap multiple benefits from surplus flower and fruit production. Extra flowers, fruit, and seeds allow parents to raise extra progeny when resources are plentiful (resource tracking), to hedge against pollination failure or damage due to predators (insurance), and to select fruit with more seeds or cross-pollinated seeds (progeny choice). And the production of surplus seeds may serve to satiate fruit and seed predators (facilitation).

Are Humans Optimistic Parents?

Humans would seem at first glance exempt from the rules of parental optimism. Anything larger than a brood of one at birth is rare, as fewer than one in twenty births give rise to twins or more. But humans are more optimistic than is immediately obvious: polyovulation is surprisingly common. It serves as a strategy to ensure against defective embryos, and it occasionally yields extra reproductive value when twins are born. Thus marginal offspring can and do serve multiple functions in humans. But that same optimism leading to polyovulation also leads to frequent brood reduction. Indeed, brood reduction is so common in humans that were we to apply the definition used by ornithologists, we should be declared obligate brood reducers. Yet we are blissfully unaware of this drama because it occurs out of sight during the earliest stages of pregnancy, especially in older mothers. I shall explore this enigmatic behavior in more detail in chapters 7 and 8.

Chapter 3

Why Parents Play Favorites

> *Now Israel (Jacob) loved Joseph more than all his children, be-*
> *cause he was the son of his old age: and he made him a coat*
> *of many colors. And when his brethren saw that their father*
> *loved him more than all his brethren, they hated him, and*
> *could not speak peaceably unto him.*
>
> —Genesis 37: 3–4

Mom Always Liked You Best

Though he was the eleventh of twelve sons, Joseph was his father's favorite. His father's gift of a coat of many colors reflected his special status as the designated heir and also, sadly, aroused the enmity of his brothers. This favoritism triggered a fratricidal conspiracy. Joseph's life was spared only when Reuben, his eldest brother, objected to the plot to murder him. Instead, Joseph was stripped of his coat, cast into an empty pit, and then sold into slavery. The brothers dipped the coat in goat blood before returning it to their father, who mourned the loss of a son he believed devoured by a wild beast. Such are the dangers of a parent playing favorites.

Parents nurture and protect their progeny, but the critical resources that parents provide—food, water living space, and warmth—are often in limited supply. This shortfall leads directly, inexorably to sibling rivalry. And such rivalries can turn fierce, especially when parents prefer some offspring to others. Why do they play favorites? For Jacob it was because Joseph was the child of his old age. But across nature, favoritism is linked to the habit of parents routinely creating more offspring than they can comfortably sustain. This is parental optimism, and a poverty of supply coupled with an excess of desire is an equation without balance. Not all incipient offspring can survive to independence, and parents compound the problem by working at submaximal levels to sustain their children. Since not all progeny can survive, parents pick the winners and losers, conferring advantages to some and handicaps to others.

The path to conflict

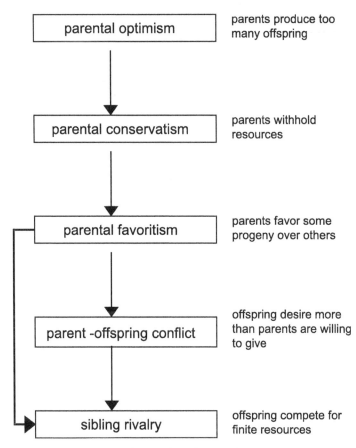

Figure 3.1. The path to conflict within the family
An initial overproduction of offspring coupled with parents that play favorites leads inexorably to sibling rivalry and parent-offspring conflict.

The Fivefold Advantage of Favoritism

Why do parents set conditions for some progeny to fail? The answer is that brood reduction is a necessary complement to parental optimism. When parents normally create more offspring than can be sustained, a mechanism for a secondary downward adjustment of clutch size is required. Mechanisms such as hatching asynchrony create phenotypic handicaps (in this case larger and smaller off-spring) that serve to make the process more efficient. In birds, brood reduction occurs even in synchronous broods, but it occurs later, meaning that more resources have been squandered on doomed off-spring. Asynchronous hatching results in earlier, less wasteful brood reduction but permanently skews resources toward the older, core sibs, and thus probably generates some unwanted and maladaptive brood reduction even when food is plentiful. This represents a potential cost of a brood-reduction strategy.

But why not just add offspring to the brood as needed? This simple and obvious strategy is thwarted by time constraints and ergonomics. Parent birds and their brood may find themselves amid an ephemeral flush of food without the ability to adjust their brood size upward in a timely manner. This would require laying and incubating additional eggs, a task incompatible with brood rearing, and even more important, it would take too long. It is much easier to begin with a brood too large and trim downward than to begin with a brood too small and revise upward.

A parent is equally related to all of his or her biological offspring. But genetic equality does not guarantee the equal treatment of offspring. The reason is disarmingly simple. Favoritism creates diversity, and diversity creates opportunity. These opportunities represent the fivefold advantage of favoritism. Favoritism allows parents to (1) make strategic investments in some progeny but not others; (2) facilitate the removal of offspring that are not needed; (3) diversify resource supplies or spread out offspring demands; (4) correct earlier parental decisions that ultimately prove faulty; and (5) reduce variation in reproductive success.

1. The Benefits and Costs of Unequal Parental Investment

> *Selfishness is not living as one wishes to live, it is asking others to live as one wishes to live.*
>
> —Oscar Wilde

An unequal allocation of resources is a consequence of playing favorites. This alone may itself be adaptive in an uncertain environment. Uncertainty surrounds the resources available for investment in offspring. One potential solution is to create a subset of core progeny cosseted from the effects of shortfall, alongside a caste of marginal offspring vulnerable to the effects of deficient resources. The marginal progeny are provisioned if resources prove sufficient but

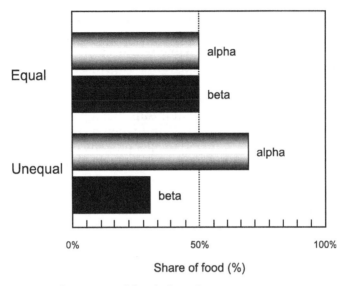

Figure 3.2. Equal vs. unequal food allocation
The horizontal bars represent food shares to a brood of two offspring (alpha and beta) under equal and unequal allocation. Under equal allocation, alpha and beta receive identical food shares. Under unequal allocation, the senior sibling, alpha, garners a greater food share at the expense of its junior sibling, beta. Such differences in food share translate into differences in expected survival rates.

allowed to perish otherwise. This of course describes the process of resource tracking. As resources become increasingly scarce, favoritism (e.g., by creation of a brood hierarchy) results in a greater fraction of resources being diverted to the core brood. Stock market analysts have a term for this: the "flight to quality," which occurs when market conditions turn adverse and capital is moved from risky to safe investments. Thus, investments are moved out of volatile stocks and into low-risk government bonds during periods of uncertainty. Parental favoritism potentially allows parents to make the flight to quality by consolidating parental investment in a smaller number of relatively high quality progeny during periods of scarcity.

Brood reduction is a potential consequence of unequal resource allocation if resources, most notably food, are limited. The bottommost members of a brood hierarchy receive the smallest food shares, which may not be sufficient to sustain them. The death of the doomed offspring, and the process of brood reduction, can be rendered more efficient if preemptive siblicide or cannibalism occurs. If and when it becomes obvious that the available resources will not sustain the full brood, further investment in the marginal progeny is imprudent. Here we encounter an application of the principle of holes: if you find yourself in one, quit digging. There is no point in squandering additional valuable resources on bad investments. Those who thought they were getting a bargain by purchasing extra Enron shares as the stock price fell obtained firsthand knowledge of the principle of holes.

A policy of unequal investment means that a parent can curb the loss of parental investment (*PI*) if and when a reduction in brood size becomes necessary. Younger, smaller progeny that are the targets of brood reduction will have consumed less *PI* at any point in the period of parental care than older, larger progeny. Investment in marginal progeny thus becomes analogous to the investment of risk capital in the stock market: one must balance the risk of losing the investment entirely against the potential gains.

Playing favorites can therefore potentially allow a more efficient allocation of resources but also bears a price: despotism. Given the opportunity, some progeny will take too much, and thus favoritism can foster unwanted and sometimes fatal sibling rivalry. Handicaps render some offspring more equal than others, and inequality creates the opportunity for selfishness. In nestling birds and broods of mammals, how parental investment is apportioned is often left to the offspring, and dominant sibs can command a disproportionate share of food.

2. Divestment of Unneeded Offspring

Some offspring ultimately become surplus to their parents' needs, and favoritism can shave the costs of elimination. One route is a one-sided sibling competition: a despotic senior sib can serve as an efficient executioner. One need look no further than obligate brood-reducing pelicans, eagles, or boobies to witness the ruthless elimination of marginal progeny by stronger sibs given a three- to seven-day head start at hatching. But once brood reduction has occurred, the decision is irrevocable. An offspring once dead tends to stay that way. Postponing the decision gains time, and time brings more information with which to make a considered decision. Playing favorites can also help parents make *better* reproductive decisions.

There is a built-in advantage to maintaining an optimistic perspective, as it is far easier to trim offspring numbers as conditions warrant than to revise upward after the fact. With the passage of hours, days or weeks, ecological conditions unfold, and provide the parent with a critical resource. No, it is not food, but something even

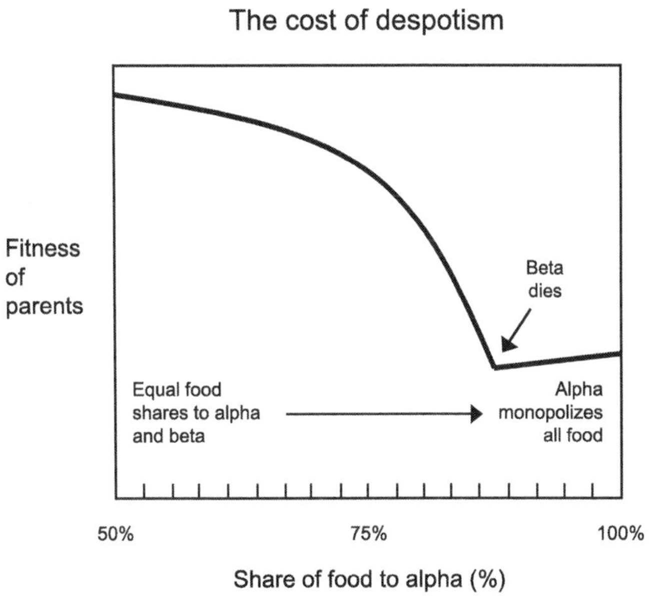

The cost of despotism

Fitness
of
parents

Beta
dies

Equal food
shares to alpha
and beta

Alpha
monopolizes
all food

50%   75%   100%

Share of food to alpha (%)

Figure 3.3. The cost of despotic allocation
When resources are plentiful, parents prefer an equal allocation of food to
all offspring. This would maximize their reproductive fitness by maximizing
the number of healthy surviving offspring. However, a despotic allocation
arises when offspring, rather than parents, control how food is shared. This
erodes parental fitness in times of plenty as the distribution becomes increasingly unequal. The figure illustrates the cost of despotic allocation in
a brood of two, with alpha and beta sharing the available food. The fitness
of parents falls as alpha's share increases from equal allocation to a complete
monopolization of all food. At some point (here a food share of 90%) beta's
food ration is insufficient for survival, and beta perishes. Beyond this point
it pays both alpha and the parents for alpha to take all available food.

more valuable: *information*. Information about the level of resources
available in the environment and, as a consequence, about the prospects for rearing a small or large family allows parents to review and
if necessary revise earlier, optimistic decisions about family size. If
resources are sufficient, a parent may opt to continue the status quo,
but if resources are found wanting, it may then become necessary to
subtract from the current clutch or brood.

Information about current ecological conditions may help parents to anticipate either chronic or acute food shortages. The behavioral response to a chronic food shortage is straightforward: brood
reduction, and the sooner the better. But acute shortfalls pose a dif-

ferent problem. The ephemeral nature of acute food shortage poses a tradeoff to the likely survivors, usually dominant siblings. They can gain food in the short run by killing off some of the competition, but doing so sacrifices a close relative and the associated evolutionary fitness. An overly hasty decision could prove expensive later if bountiful food conditions return soon enough. Delays in the completion of brood reduction may diminish the risk of an error, and might be favored by natural selection if the burden of an acute shortfall can be pushed onto the junior sib. Establishing and maintaining dominance—ensuring that marginal sibs receive a smaller food share—are means to that end.

But what defines an acute, as opposed to a chronic shortage? In a broad sense, a shortfall occurs whenever demands exceed supplies. But what is the appropriate time scale? Are brief hunger pangs a sufficient incentive to rid oneself of a sibling rival for limited food? Or does one require better reason than that? The answer is far from simple and revolves around the nature of variation. Some variation is predictable. One can be assured, for example, that more food will be available during summer than in winter. But some variation cannot be foreseen. And variation in both supplies and demands for food surely occurs at many levels, ranging from decades or longer for slowly changing oceanographic or climatic conditions to minutes or seconds for rapid changes in weather.

One can usefully define acute shortfalls as those occurring *within* a breeding episode, with food supplies temporarily dipping below current demands, and chronic shortfalls as those extending over an *entire* breeding period, with average food supplies failing to meet average demands. With a chronic food shortage, the need for brood reduction is obvious. But what about acute food shortages? Might parents or offspring sometimes be acting too hastily, forfeiting the lives of siblings or offspring when it was not really necessary? The answer rests on the long-term costs of inaction (e.g., depressed growth, elevated mortality) and whether favorable food conditions are likely to return soon.

The matter of food shortfall is also tied to parental choice. Both acute and chronic food shortages can arise even when parents are capable of providing sufficient food for the brood but prudently choose not to. If greater effort jeopardizes parental prospects for survival and future reproduction, it may be politic to withhold further, potentially costly investment in current offspring. And rearing conditions that vary widely (e.g., spectacularly unpredictable foraging success) favor strategies that hedge against uncertainty. One option is to hoard food in caches, or as large fat stores. Or failing that, another is to use parental restraint as a hedge against runs of bad luck that jeopardize the survival of a larger brood. Thus, surplus work capacity may provide insurance against acute—but not chronic—food shortages. By playing favorites, parents can hold the option of cheap brood reduction in reserve.

3. Benefits of Diversification

Playing favorites can allow parents to diversify either the needs of offspring or the resources available to sustain them. Offspring in a synchronous brood or litter will reach the point when they make their peak demands simultaneously, resulting in a maximum load on parents. This problem may be compounded if sibling competition is intensified in broods of equally matched progeny, elevating their demands even further. One solution is to spread out the demands by staggering birth or hatching. One need look no further than human twins for an obvious example. Even today in Western societies with access to maximum nutrition and high-quality health care, breast-feeding twins (or higher multiples) place a heavy burden on a mother. In preindustrial and rural societies, births of twins routinely produced fewer surviving progeny than singleton births (particularly if both infants were more expensive sons), and rates of maternal mortality were sharply higher. Humans are better off in most cases with one baby at a time.

The benefit of spreading out peak demands also finds another human parallel. Sociologists have long known that larger family size is associated with lower cognitive performance in children (as measured on a variety of standardized tests) and with reduced educational attainment that ultimately affects career success. Educational attainment is a good variable to study because it is easy to measure. How many years of grade school did a child complete, or how many years of college or university?

Why does family size have such a strong effect on educational performance? Two hypotheses have garnered the most attention. Under the *confluence* hypothesis, a child's intellectual development is a function of the intellectual milieu of the family. A single child interacting with two parents receives the maximum stimulation. As family size grows, children spend more time with other children, lowering the average intellectual climate, the notion being that a child learns more and faster from adults than from other children. The confluence hypothesis initially garnered considerable support, as it seemed to explain many of the observed patterns concerning family size and education attainment. But it has come under increasing criticism in recent years for what it cannot explain, such as the lack of difference between singleton children and firstborns in two-child families.

The alternative is the *resource dilution* hypothesis. Similar in some respects to the confluence hypothesis, it argues that parentally provided resources become more thinly spread in larger families. Behavioral ecologists would describe this as a natural outcome of the principle of allocation (see chapter 2). Thus in larger families the key resources, such as the time parents spend with individual children, or family income, are diluted; it is easier to pay for one child to go to college than five. The phenomenon is well studied: Access to items that might influence educational performance—a computer in the home—is lessened in larger families. Children in big families spend less time reading and more time watching television. And so on. The patterns hold even when the potentially confounding inverse relationship between family size and family income is removed statistically, though not as strongly.

The resource dilution hypothesis has gained considerable support, though there are still some unresolved patterns. Under resource dilution singleton children should fare best, but they seem to do no better than firstborns in two-child families. If an outsider might offer a friendly observation, this might illustrate the law of diminishing returns at work, or simply that resources are not markedly strained until the number of children exceeds the number of parents. That notwithstanding, sociologists have noted that children in families with longer intervals between births seem to fare better in terms of the educational performance of the children. Here the principle of diversification would seem to hold, and indeed there is evidence of this. Longer birth intervals potentially allow families to replenish expended resources, easing the burden of caring for a large family.

In avian studies, support for the hypothesis of peak load reduction has been mixed, owing in part to the difficulty in measuring the key variables. How does one measure workload? How does one measure family success? There appear to be savings from spreading the brood out by asynchronous hatching, but whether these are significant (i.e., affect fitness) is unclear. Unbeknownst to ornithologists, sociologists have provided strong support for the benefits of peak load reduction, but in humans rather than birds. Here both groups, sociologists and biologists, would appear to have much to learn from each other, but neither side seems to have tried very hard.

4. Correcting Earlier Decisions

Parental favoritism may also serve to correct errors that become evident ex post facto. Staggered hatching or birth asynchrony may impose an unnecessary burden on marginal progeny if resources later prove sufficient for all. In a stochastic world—one with imperfect information about forthcoming events—mistakes will be made. Sometimes it may be possible to correct such mistakes after the fact. And some parents seem to do this. They correct results of earlier manipulations, such as a hierarchy of large and small offspring, by preferentially feeding the smallest offspring in the brood. This is particularly common in birds with biparental care.

5. Bet Hedging and Brood Reduction

A quirk of evolutionary mathematics can yield a further advantage to playing favorites. We normally think that producing the most surviving progeny maximizes evolutionary fitness. But that is not quite true. It matters how these offspring are produced, and strategies that reduce variation in reproductive success will, all else being equal (and sometimes even when all else is not equal), be favored by natural selection. The process involved is bet hedging.

Reduced variation in reproductive success can be advantageous independent of whether the average success (the arithmetic mean) is increased, and can even be selected when the average success declines, though within limits. In essence, variability in reproductive success can be traded off against average reproductive success. In environments that vary across time, the geometric mean (defined mathematically as the nth root of the product of n numbers) of lifetime reproductive success is a better measure of fitness than the arithmetic mean.

When is geometric-mean reproductive success likely to be important? Consider the simple case in which reproductive success varies from year to year (as might happen, e.g., due to large-scale fluctuations in climate or oceanographic conditions) and such effects fall on all members of the population equally. Now imagine that there are two genotypes (organisms with a particular genetic makeup) in the population, and to keep things simple, organisms of either type all die at the end of the year. The first genotype is conservative and produces only a single offspring each year, independent of whether breeding conditions are good or bad. The second genotype is a risk taker and lays four eggs. When conditions are good, it produces four offspring successfully, but when conditions are bad, due to overcrowding it produces none at all. If good and bad conditions are equally likely, then our risk-taking genotype averages two offspring per year. Contrast that to our conservative genotype, which always produces a single offspring.

The average rate of reproduction for the risk taker is double that of the risk avoider. But look at what happens in the first bad year. The risk-taking genotype fails to produce any offspring and, because we are talking about an annual species, goes extinct. The risk-averse genotype produces only half as many offspring on average but does not go extinct when conditions turn bad. Here arithmetic-mean reproductive success does not predict evolutionary success. Rather, the genotype with the lower—in this case zero—variation in reproductive success fares better. This example is an oversimplification but captures the essence of the problem. Reducing variance in reproductive success can be enough to succeed in the evolutionary game.

Here the appropriate measure of fitness is the geometric mean of reproductive success each generation. The geometric mean of our risk taker is the square root of 0×4, which is equal to zero. The geometric mean of our risk-averse genotype is the square root of 1×1, which is equal to one. As the geometric mean falls with increasing variability, a genotype reducing its variance can be favored even if it sacrifices average (arithmetic mean) reproductive success.

Strategies that reduce variance at the expense of average reproductive success are referred to as bet hedging. Bet hedging will be most important in short-lived organisms, as repeated breeding (iteroparity) is itself a variance-reduction mechanism. Members of a species that breed for one hundred consecutive years will encounter both good and bad conditions over their long life span. Their genotype is not driven to extinction by an occasional adverse year, even if they adopt a risk-prone reproductive strategy. But for species whose members live only one or two years and breed only once or twice, occasional bad years can have disastrous consequences. A brood reduction strategy—based on parental favoritism—is a strong candidate for bet hedging because it tends to reduce variation in reproductive success. It does so by ensuring a smaller number of higher-quality offspring than the alternative of whole-brood survival, under which offspring quality fluctuates widely. And it does this by ensuring that food is not shared equally among progeny— that is, by playing favorites.

So much for *why* parents play favorites. In the next chapter I examine how.

Chapter 4

How Parents Play Favorites

> As a free-trading Malthusian he also hated primogeniture,
> that is inheritance by the oldest male; this had to be destroyed
> "to lessen the difference in land wealth & make more small
> freeholders." It would create more competition and sift out the
> clever, ingenious, "fittest" sons.
>
> —Adrian Desmond and James Moore, *Darwin:*
> *The Life of a Tormented Evolutionist*

HUMAN PROGENY are not exempt from the rules of parental optimism, and primogeniture is a clear example of how humans play favorites. With the family property passed to the eldest son, older and younger progeny are rendered decidedly unequal, even though from the parents' perspective the offspring are on average genetic equals. Ceteris paribus they are equally worthy of investment. But all else is not equal. In a vast array of species, parental—and usually maternal—manipulations of phenotype (the morphology, physiology, and behavior of an individual) place some offspring at a disadvantage at the very start of life.

What Is a Phenotypic Handicap?

A phenotypic handicap arises when a parental manipulation impairs an individual's competitive status among its brood mates. Such manipulations come in two broad types. The first holds the phenotype constant but renders offspring unequal by staggering their development. Hatching or birth asynchrony provides a developmental head start to some progeny and a delay to others. This may yield a commanding advantage to an individual that is otherwise identical to its sibling, simply because it is larger and/or develops motor skills sooner. At the extreme—in parasitoid wasps—even a few minutes' difference in when an egg is laid can spell the difference between life and death for the offspring. Humans are excellent examples of this sort of handicap. Instead of giving birth to a litter of six, seven, or eight, as do rabbits, mice, and pigs, human mothers typically give birth to singleton offspring, occasionally two. And infant mortality, when it occurs, is typically higher among later-born offspring.

The second type of handicap involves modification of the offspring's phenotype: some eggs or neonates may be larger than others or more pugnacious because they received hormonal boosts from their mother. This type of handicap can generate competitive asymmetries between two (or more) offspring born or hatched at the same time. Of course, both types of handicaps—staggered development and modified phenotype—can and do occur together. Some birds for example appear to diminish or exaggerate the competitive asymmetry among offspring created by hatching asynchrony by adjusting egg size. Large last-laid eggs tend to reduce the penalty of hatching last, whereas small eggs exaggerate the effect.

Birds are models for the study of the effects of phenotypic handicaps on sibling competition for a simple and obvious reason. We can watch them from the time of hatching to when they leave the nest, and sometimes beyond. Many birds build open nests in accessible locations, making them much more observable than, say, mammals that gestate offspring inside a womb and afterward care for the progeny in inaccessible nests, dens, or even pouches. Contrast this to an egret's or blackbird's or robin's nest where a video camera can easily be set up to study nestling behavior and development. For this reason, we know a great deal about how parent birds play favorites.

How Birds Play Favorites

The primary mechanism of favoritism in birds is hatching asynchrony. Parent birds often stagger the hatching intervals of chicks, and in doing so confer lethal handicaps on last-hatched offspring that differ from their elder nest mates mainly in that they are last to the dinner table. Superimposed on this basic mechanism are secondary manipulations such as differences in eggs that translate into differences in the size of hatchlings and the food stores they carry as yolk reserves. Some parents even play chemical games, dosing certain nestlings with extra quantities of steroid hormones that render them especially belligerent. Recent work shows that these handicaps render the chicks decidedly unequal. Let us now examine the methods of favoritism in birds more closely.

Primary Versus Secondary Handicaps

Age differences at hatching translate directly into competitive asymmetries as older, larger, and stronger chicks develop motor skills earlier and routinely outcompete their younger nestmates for food. The effect of hatching asynchrony is normally so powerful as to over-whelm the secondary manipulations of egg size and hormone level. Hatching first is almost always more important than hatching from a larger egg, though larger egg size may partially close the competitive gap. Recent work on hormonal bias among nestlings suggests that mothers may modulate competitive asymmetries within broods by providing some progeny with a testosterone boost. Last-hatched canary chicks beg harder at least in part because the mother provides a higher dose of testosterone to the last-laid egg. These later-hatched chicks fare better in early nestling competition, though the effect appears to be relatively short-lived.

Female birds deposit hormones in their eggs in varied patterns. Cattle egrets and zebra finches put more testosterone in first-laid eggs, and the amount decreases successively with later-laid eggs. The opposite pattern occurs in canaries, American kestrels, and red-winged blackbirds: the females add successively more testosterone to later-laid eggs. Zebra finch females also add extra steroids to all eggs when they mate with more attractive males, while still retaining the within-clutch pattern. The story grows more complex as there are also costs—such as depressed immune systems—associated with high doses of hormones, and growing evidence shows that parents also tinker with the embryonic immune systems as well.

But the effects of hormonal favoritism seem to be overshadowed by hatching asynchrony, suggesting that hatching asynchrony is the primary handicap and that differences in egg/neonate size or hormonal titer are mechanisms to modulate the competitive gap caused by asynchrony. These secondary handicaps presumably assume greater import, though, when chicks hatch synchronously.

Parent birds thus create surplus progeny and then play favorites within the brood under a strategy of parental optimism. Rank within the brood becomes a powerful determinant of offspring behavior, in part because they are assigned different roles by parents and in part because offspring of different ranks face differing prospects.

How Blackbirds Play Favorites

By handicapping certain of their brood, parents keep the costs of maintaining the potential surplus of offspring low. Where key resources—particularly food—are distributed according to the outcome of sibling competition, marginal offspring are likely to get less than their fair share. But under resource tracking, this penalty falls as food supplies rise. This simple notion has been surprisingly difficult to test. But we do see evidence for it in the avian equivalent of drosophila, the red-winged blackbird.

Red-winged blackbirds clearly use marginal chicks to track food supplies from year to year. In southern Manitoba most female red-wings lay four eggs, with two combinations of core and marginal offspring most common: either two first-hatched core offspring and two later-hatched marginal offspring, or three core and one marginal. The mean number of hatchlings varies little across years, averaging about 3.6 nestlings. But the average number of nestlings that survive to leave the nest varies by more than 50% across years, from a low of just over 2 to a high of 3 1/4. The number of core nestlings that survive to fledge is almost fixed, and is very close to 2 every year. Nearly all core chicks survive nearly always. Rather, the variation in family size stems from variation in the survival of marginal chicks, which ranges from near zero in the worst years to over 80% in the best years. This variation is attributable to variation in growing conditions for the young. Survival is highest when the spring and early summer is warm, and lowest when cool. The redwing reproductive strategy can be described quite simply: when the going gets tough, the marginal offspring perish.

Reversible Handicaps

Hatching asynchrony is the key phenotypic handicap in this system. It serves to keep the cost of maintaining the "surplus" marginal offspring low when they are not needed, but the handicap is not so great that it cannot be reversed under good brood-rearing conditions. The marginal offspring are surplus in only some nests and some years but not others. Experimentally adding or removing marginal nestlings to blackbird clutches has almost no effect on the prospects for core nestlings, but the reverse is not true. Adding or removing core nestlings has a dramatic effect on marginal nestlings. Remove a core nestling, and the growth and survival prospects of a marginal nestling soar. Add a core nestling, and the prospects for the marginal nestling dim.

This asymmetric sibling rivalry highlights the second key function of marginal offspring: replacement. Roughly one in ten blackbird eggs will fail to hatch, and some of these will be core eggs. The survival of marginal progeny jumps in clutches with hatching failure, as the presence of the marginal egg serves as insurance against core failure. By playing favorites, blackbird parents keep a reserve of marginal offspring that can be maintained at low cost but can also be drafted into use if one or more blue-chip members of the brood fail to make it past the starting gate, and/or breeding conditions are unexpectedly favorable. And since breeding conditions depend on weather, they are effectively unpredictable more than a few days in advance. By hatching their broods asynchronously, blackbird parents can have their cake (or perhaps dragonfly) and eat it too.

The key feature of the system is the phenotypic handicap. It buffers core offspring from the effects of overcrowding but allows parents to maintain a caste of low-cost and easily disposable offspring as a hedge against uncertain ecology or development. For such a system to work, the handicap must be reversible. And it is. Marginal offspring can and do thrive when conditions are favorable.

Though brood reduction has gained the greatest attention from ornithologists, beginning with David Lack, it is not the exclusive province of birds. In recent years field biologists have uncovered parallel phenomena in a diverse array of organisms, often without realizing it, as the lines of communication across fields are frequently sparing. Among these organisms are bizarre little mammals that dwell in the forests of Australia.

How Marsupials Play Favorites

Brood reduction of course is not just an avian habit, though it is most easily studied in birds. And brood reduction is not just a mechanism to track varying food supplies. It is, in fact, a general-purpose mechanism that can be used for various ends. And one of these is the adaptive adjustment of a brood's sex ratio. In birds and mammals, sex allocation is governed by a Mendelian lottery, the random allotment of chromosomes to gametes (male or female germ cells, e.g., sperm or egg). In mammals, it is the male that "decides" the sex of the offspring, as it is the male that is heterogametic (i.e., carries a female X and a male Y chromosome). In birds it is the reverse, and females are heterogametic. Under a fair Mendelian lottery there should be an equal number of male and female gametes produced by the heterogametic sex (males in mammals, females in birds), and hence an equal chance of male or female offspring at conception. But there are occasions when it would pay to prefer one sex to the other—if there is a temporary anomalous preponderance of one sex in the population, for example, or one sex is larger and more expensive to rear, or one sex is more likely to become a future competitor with a parent. Here the Mendelian lottery gets in the way of the adaptive adjustment of sex ratio. Parental optimism allows a potential solution. Create more incipient offspring than needed (with an equal sex ratio at conception), and tailor the brood to current needs via brood reduction. Certain marsupials (pouch-bearing mammals) do exactly that.

Australian antechinuses are extraordinary animals. They live in the forests and woodlands of eastern Australia and look to the casual observer something like a mouse, but experts say they more closely resemble a shrew. In fact they are marsupials (though their "pouch" is more of a cup than a closed sac) and not at all closely related to rodents that are placental mammals. Adult agile antechinuses are small, ranging from 20 to 40 g, with males larger. The offspring are smaller, astonishingly so. At 1/16 gram at birth, they are the smallest babies of any mammal. And they do not live long, males less than a year and females up to three years. Breeding occurs once yearly, and all the females in a population give birth at the same time. The litter size varies, but at birth the number of babies usually exceeds the number of nipples and often by a wide margin. This is bad news for those unlucky enough not to find a nipple in the neonatal game of musical chairs, as a baby requires sole access to a teat for the first month. Losers in the game forfeit their lives. But perhaps what is most remarkable about *Antechinus agilis* (the agile antechinus) is that the males are semelparous. That is, like Pacific salmon and bamboo, they breed once and die. Antechinuses and their close marsupial relatives are the only known semelparous mammals. Semelparity means that you need not hold anything back for reproduction. Pacific salmon make exhausting spawning migrations from salt to fresh water, journeys on which they do not feed. Rather, they use their stored energy to fuel their journey to the breeding grounds where they spawn and die. Such a one-way trip means that female Pacific salmon can build more and larger eggs than their iteroparous (many-time breeding) Atlantic salmon counterparts.

Male antechinuses give their all during breeding, and like Pacific salmon die en masse. Thus, for a short period each year, the adult population of agile antechinuses is all female. The fact that males are semelparous, and females potentially iteroparous, has dramatic consequences for the breeding strategy. Female antechinuses live on a home range that they often share with their daughters but not sons: males are summarily evicted shortly after weaning. Daughters but not sons are potential competitors, probably most importantly for food during lactation. Though the babies are born tiny, they do not stay that way, and at the end of the three-month nursing period the mother is sustaining a brood that may be five times her body weight. Mothers then become selective about whether they carry sons or daughters. A yearling female, who may live to breed again, prefers a brood with more sons. An old female, who will not live to breed again, produces an all-female brood. Work by Andrew Cockburn and his students have shown that this adjustment is done postnatally by sex-selective brood reduction. Again, the basic rule of parental optimism is applied. Create more incipient offspring than will be sustained, and trim the excess cheaply to tailor the brood to your needs. Here females that will not live to see another breeding season are unconcerned about potential future competitors: the future is now. Yearling females are more sensitive to future competition and facultatively adjust the sex ratio of their brood.

This, however, is not the whole story with the agile antechinus. In some populations females are nearly semelparous: most perish before they can breed a second time. In these populations, females give birth to litters that are five-eighths female. We just don't know how female antechinuses do it, though the mechanism appears to operate very early. Most likely female antechinuses are able to bias fertilization toward female-bearing gametes via an as yet unidentified mechanism. For evolutionary biologists there is always something deeply disturbing about prefertilization sex ratio biases, as they imply that the Mendelian lottery that governs sexual reproduction is rigged, which challenges fundamental assumptions of evolutionary biology. I imagine that the introduction of quantum theory produced the same queasy feelings among classical physicists.

Brood Reduction in Rabbits

Littermates compete for their mother's milk in other young mammals as well. And as with nestling birds, success in this sibling competition can spell the difference between life and death. Litters of European rabbits average seven to nine, though brood size exceptionally reaches eighteen. As brood size rises, so too does competition for milk, especially since mother has only six to eight nipples, and so too the likelihood of litter reduction. With more pups than nipples, sharing must occur if all are to survive, and nipple switching is the norm. Rabbit mothers are not doting parents. They cache their babies in an underground nest and return at daily intervals for only three to four minutes to feed their pups. Mother secretes a pheromone on her belly to help pups locate nipples quickly, and the problem of overcrowding is solved by nipple switching. Each pup feeds at a nipple for ten to twenty seconds and then moves on. Runts, however, fare poorly in this scramble for food, and pups underweight at birth are more likely to succumb to starvation. The death of a pup liberates its food share for the survivors, who benefit from an increased per capita consumption.

Brood reduction in rabbits is not just a simple story of sibling competition for food. Rabbit pups also consociate to conserve warmth by huddling, and the energy savings can be substantial. An individual can shave up to 40% from its metabolic expenditures and divert the savings to other purposes such as growth. What outwardly appears to be a simple sibling competition among rabbits is in fact a layered dynamic. Litter mates both compete and cooperate simultaneously, taking from and contributing to an energy pool that sustains all offspring. Rivals are also partners on the journey from embryo to independence.

Sibling rivalry is not the only means of trimming brood size to affordable levels. When food supplies are restricted, house mouse dams cannibalize offspring to reduce brood size; deer mouse mothers, by contrast, are content to rear a full brood of malnourished progeny.

How Plants Play Favorites

Plants are also optimistic parents. Many species routinely produce mature fruit from only a fraction of their female flowers, aborting the surplus. They use the initial surplus to track unpredictable nutrient and pollen supplies, as a hedge against seed predation, and to select high-quality progeny. They abort damaged seeds or seeds with chromosomal defects, or eliminate seeds that are the product of self-pollination in favor of cross-pollinated seeds. Plants, just as birds, thus use surplus offspring for the purposes of resource tracking, insurance, and progeny choice.

In many respects plants are model systems for the study of parent-offspring relations. They are easily accessible, and easily manipulated. While it is relatively difficult to manipulate the food supplies of wild birds or mammals, one can do so quite easily with plants by clipping leaves (defoliation reduces nutrient production) or adding fertilizers. In fruit trees a diminished leaf area increases the rate of fruit abortion; conversely, adding fertilizer during the flowering period greatly reduces fruit drop. Plants adjust their fruit set to the level of available resources.

Plants also practice favoritism within their "brood"—some herbaceous species mix normal-size (core) with smaller (marginal) fruit that lag behind in growth and are usually aborted. But when the "core" fruits are thinned, the marginals grow to maturity. This conditional response is the signature of a replacement strategy.

Plants also use selective fruit abortion to practice quality control among their progeny, and in particular to favor cross-pollinated seeds over self-pollinated seeds. We need look no further than the familiar apple tree, which preferentially aborts fruits from self-pollinated flowers over cross-pollinated flowers. But when fertilizer is applied, more fruits from self-pollinated flowers are matured. Here there are two castes of seeds: the low-quality "marginal seeds" that result from self-pollination, and the high-quality "core seeds" that result from cross-pollination. When resources are plentiful, more of the surplus marginal seeds from self-pollinated fruit are matured and provide extra reproductive value. As in birds, so too in plants.

Different Species, Same Idea

Birds are very different in obvious ways from mammals, and especially humans. Can we then draw any useful parallels between the two? Instead of trying to make my general case for the affirmative here (I shall use the remainder of the book to do that), let me examine one case in which birds provide a useful experimental model for understanding human families: the adaptive significance of brood reduction.

Brood reduction is a routine feature of the strategy of parental optimism. In birds, marginal offspring are the last to hatch and, when brood reduction occurs, the first to die. In humans, the "brood" consists of a series of sequentially produced offspring whose periods of dependency normally overlap. Infant mortality—particularly in rural societies—is sensitive to the birth intervals, being highest when short, for example, two babies born less than eighteen months apart. It is the younger infant who is at greater risk, though mortality among older infants also rises with short birth intervals.

Ornithologists have long asked whether brood reduction per se is adaptive or arises as a side consequence of other reproductive considerations, such as selection for a rapid completion of reproduction when nests are vulnerable to predators. Adaptive brood reduction should provide relief for the surviving family members. If it arises chiefly from other selective forces its occurrence may well be maladaptive. We can pose the same question for infant mortality in humans. Is infant loss adaptive? Does it provide relief for the mother and surviving infant? The answers to the same question posed for two very different taxa—birds and humans—are eerily parallel.

Some ornithologists have questioned whether avian brood reduction is indeed adaptive, noting that broods in which it has occurred tend to show lower growth rates even *afterward*. Surely brood reduction should provide relief for the remaining brood members, and the growth rates of surviving nestlings should rise to those for avian families in which brood reduction did not occur. Essentially the same argument has been made concerning infant mortality in humans. Among families in which one infant has died the older child still remains at an elevated risk of mortality. Surely then, brood reduction/infant mortality has not provided relief, and hence cannot be adaptive.

This logic is appealing but ultimately specious. Brood reduction/infant mortality does not remove the underlying problem: inadequate parental resources to sustain the family. But is it adaptive? The question is not whether the loss of one (or more) offspring produces equality with other families in the same population. The necessary and sufficient condition for it to be adaptive is that brood reduction provides relief from the overcrowding that would otherwise have occurred. That is, the parents would do better with brood reduction than without. And here the answer is clear: without the reduction, things would have gotten much, much worse. And this follows directly from the principle of allocation.

The principle of allocation naturally holds for humans. Most people who purchase a new car—say a Lada SUV—take out a car loan. But some people, either because of bad luck or financial mismanagement (the latter being particularly likely for those who purchase a Lada SUV), will default on the car loan. Are people who default on the car loan more likely to default on their house payments too? Very probably. But by shedding the burden of car payments, our homeowner may now be able to meet the mortgage payments and avoid losing the house too. The underlying problem is that supply (bank balance) does not meet demand (bill payments due). If, however, the individual did not default on the car loan, things would grow much, much worse very quickly, and almost certainly that individual would default on both the car and the mortgage payments in the near future, and be searching for an unoccupied cardboard box in which to set up residence.

Humans Play Favorites Too

The same general principles apply to the economics of raising avian and human families. Ethical considerations clearly prevent the obvious experiment—replacing the victim of brood reduction—from being conducted in humans. But such an experiment can be and has been done in birds. The results are clear. Replacing the victims of brood reduction places the remaining siblings at an elevated risk of reduced growth and higher mortality. The benefits of brood reduction thus are prophylactic: the loss of an offspring now avoids greater ruin in the future. That result holds for birds. Does it also hold for humans?

The key question is why brood reduction afflicts some families and not others. In birds, the lower growth rates provide important clues. Food shortfall is the root cause perhaps because some parents are less skilled at foraging or in poor health or are just plain unlucky. In humans, the problem is more complex. High infant mortality is associated with short birth intervals. But why? Some have speculated that babies too close together impose a heavy burden on the mother, leaving her physiologically exhausted. But the data do not bear this out. At present, the cause of elevated infant mortality in families with short birth intervals is something of a mystery. But we may be focusing on the wrong variable. As with birds, we may be looking at symptoms, not causes.

Breast-feeding provides a natural contraceptive. While nursing, a mother produces the hormone prolactin, which in turn suppresses ovulation. In rural populations where babies are exclusively breast-fed, this "lactational anovulation" results in birth intervals that average about thirty-six months. But this does not hold for mothers in poor health. Mothers in poor physiological condition and/or afflicted by parasites or pathogens cannot sustain breast-feeding. This has two key effects. First, the mother resumes ovulation and is likely to become pregnant sooner. Second, it places the health of both the infant at hand and the infant to be at risk.

Thus short birth intervals are a red herring. It is not the short birth interval per se that causes the elevated mortality of infants, but rather it is what causes the short birth interval that also causes infant mortality. Every introductory statistics course has an obligatory lecture about correlation not being causation, and here we have a textbook case. Just as in birds, brood reduction in humans is a symptom of an underlying problem, not a cure. Thus is brood reduction adaptive in cases of infant mortality? Insofar as it prevents a bad situation from getting worse, playing favorites in human families probably is adaptive.

Birth Order and Favoritism

Primogeniture arises where land is critical to economic success and in short supply, enhancing the survival odds of at least one line of the family across generations. Parental favoritism thus encourages diversity within the family: if the eldest son is to receive the farm, younger siblings need to find other occupations. But primogeniture is only one form of parental favoritism in humans. I shall explore others in later chapters. Playing favorites has far-reaching consequences, and one of these is that it sets the stage for conflict between siblings, and between parents and offspring. That is the topic I shall turn to next.

Chapter 5

Family Conflict

Genetic Conflict between Parents and Offspring

At age thirty-nine Charlotte Brontë found herself the final survivor of six children. Tragedy had hung gloomily over the Brontë clan, but now in the new year of 1855, newly married and pregnant with her first child, she might look forward with some optimism. Looking back was surely painful, as the family that had given rise to the most famous trio of siblings in literary history was plagued by misfortune. Her mother had died when Charlotte was only five, followed shortly after by the deaths in childhood of her two older sisters. Her brother Branwell, the model for the character of Heathcliff in Emily's *Wuthering Heights*, died after a short unhappy life in September of 1848. Emily succumbed to tuberculosis in December of the same year, and Anne the following May.

Charlotte was best known for her dramatic romance novel *Jane Eyre* but had not enjoyed equal success in her real life. After turning down several earlier offers of marriage, she accepted the proposal of the Reverend A. B. Nichols and was married in June of 1854. Sadly, Charlotte was not to escape the dreadful fate of all the others: the hope of a new year soon gave way to despair, and she began a slow and agonizing descent into oblivion. The onset of pregnancy also brought the onset of persistent nausea and vomiting. She was unable to eat and soon bedridden, growing ever weaker. By late March the relentless sickness had exacted its toll—she became delirious and was soon dead. The technical cause of death was hyperemesis gravidarum, an extreme and debilitating form of pregnancy sickness, but in truth she fell victim to parent-offspring conflict.

Conflict between the generations is eternal. Mothers and daughters, fathers and sons seemingly differ over everything. Should brother and sister be left to decide how much pie each should get a squabble is sure to ensue, each desiring a greater share than the other is willing to give, and the parentally arbitrated decision of equal shares leaves neither happy. The squabbles, the pleading the skirmishes between siblings, the quarrels between parents and offspring, are a normal part of family relations. But why? Why should parents and offspring differ in their outlook on life? Offspring rely on parents over the period of parental care for food, protection, guidance, love, and indeed for their very survival. So why defy those who offer so much?

Rebellion is a natural part of childhood, and the notion that there is a bit of genetic programming lurking underneath parent-offspring relations is not far-fetched. Behavioral ecologists expect parents and offspring to disagree often: over how much food to provide, over how many children to have, over when the kids should leave the nest. And disagree they do. Offspring programming says this: Haggle, plead, beg, and negotiate for more. A bit more food, a bit more love, a bit more attention. Exaggerate when it serves your own interests, and probe to find your parents' limits. Then step beyond them. At the level of genes greed can indeed be good. If you do not secure that extra bit of parental care it is likely that someone else will, someone less related and less valuable to you. You and your genes are only an offspring once, so make the most of the opportunity.

The parents' programming delivers a different message: Resist offspring demands. Be frugal with limited resources. Do not believe everything the kids are telling you, because they are likely to exaggerate. Do not reward selfish behavior unduly. If you are a parent bird or mammal you are likely to breed not once but many times; thus your programming instructs you to take the long view. A parent must be thrifty with limited resources, making sure to set something aside for future progeny. Clearly the programmed messages for parents and offspring are different and often contrary. One party is engaged in a reproductive marathon whereas the other has a shorter distance to cover and desires to travel faster. For offspring the future is now.

The majority of mammals today are placental; the exceptions are those weird things with pouches (marsupials) and truly bizarre mammals that lay eggs (monotremes). In placental mammals mothers invest enormous sums of time and energy on pregnancy and lactation. Every embryo carried to birth is costly, and mothers have only a finite supply of resources to spend. A mother can invest all her resources on current offspring in a single, exhaustive effort or defer breeding entirely, saving all her resources for future offspring, or some of both. The law of diminishing returns holds not only for economics but for reproduction too. More is better, but with declining returns. The payoff for ever higher investment in current offspring will grow ever smaller, and at some point it is better to hold back for the sake of future reproduction. Mother must find the correct balance between current and future offspring. That balance, however, is likely to be contested by the offspring. This is the basis of what the evolutionary biologist Robert Trivers called parent-offspring conflict.

Parent-Offspring Conflict

Imagine a gene that resides in an offspring. If future brothers and sisters are full sibs, the probability that each of them will share a copy of the same gene will be 50% (a probability that falls to 25% for half sibs). If that same gene is sitting in one of the parents, the likelihood that any of its offspring carries a copy is 50%. Thus, from the offspring point of view, future offspring are only half as likely to share a copy of a gene that resides in the current offspring; whereas from the parental point of view all offspring are equally likely to carry that same gene. One way of looking at it is to say that an offspring is twice as related to itself as it is to its sibs, whereas a parent is equally related to all its offspring.

Now imagine that a gene can tell whether it came from mom or dad. And let us also imagine a polygamous mating system, so that mom chooses a different partner for each batch of children. Things now get very interesting. A gene from dad in an offspring (a paternal gene) has a different perspective on life, the universe, and everything than a gene from mom (a maternal gene). Since this mother will not be involved with any future paternal genes, the gene from dad will benefit by taking from mom everything it can get: the fate of that gene, and copies of that gene to be found in relatives, does not depend on its mother beyond the current bout of reproduction. Thus in the colorful language of gene-speak, it does not care whether mom gets run down or even dies after the period of parental care is over.

A gene from mom has an entirely different perspective on life. Mom expects to give birth to future offspring each of which has a 50% chance of carrying the same maternal gene. This particular gene has a vested interest in mother's future reproduction, whereas the paternal gene does not. In such cases we expect paternal genes to behave more selfishly than maternal genes. And guess what. They do!

We can see this effect clearly in the differing contributions of paternal and maternal genes to pregnancy. Paternal genes are chiefly responsible for constructing the placenta, an invasive organ that is designed to gain access to maternal resources. Maternal genes show a greater involvement with the construction of the embryo itself. We can see this in rare cases when developmental accidents lead to an embryo receiving two copies of chromosomes (either portions of chromosomes or entire chromosomes) from one parent and none from the other. Even more extreme cases arise when two copies of the entire genome—the complete set of chromosomes—come from one parent. A hydatidiform mole results if two copies of the paternal genome are present. This kind of mole is not a small myopic mammal but a placenta that grows wildly with a poorly developed or absent embryo.

A teratoma results if two copies of the maternal genome are present. It is a scrambled mass of tissue in which we might find hair and teeth and other bits and pieces of an embryo, but in all the wrong places. While usually benign, teratomas can sometimes be dangerous to mothers because they can start growing spontaneously in a mother's ovary, giving rise to ovarian cancer. Until the late 1980s evolutionary biologists thought it did not matter whether a gene was inherited from mom or dad, but we were wrong. It matters a great deal.

An imprint is a mark on a gene that says whether it comes from mom or dad. More important, the imprint disables that particular gene, meaning that there will be only one functioning copy of the gene (the copy inherited from the other parent) instead of the normal two. This in itself is puzzling to evolutionary biologists. Diploidy is the normal situation in sexually reproducing organisms. It means that each offspring inherits two copies, not necessarily identical, of every gene. The advantage of diploidy is that a backup copy is available if one gene is rendered nonfunctional by mutation. This is an important built-in redundancy—a fail-safe mechanism—that we all enjoy. The puzzle is that imprinted genes shed their natural insurance policy. Gene pairs with imprinted copies are more vulnerable to being deactivated by mutation, as only one copy of the gene is active. This perhaps helps to explain the relative rarity of imprinted genes: about four dozen are currently known in humans, out of the roughly twenty-five thousand genes that comprise the entire human genome. Although rare, imprinted genes have important effects, and we see their handiwork during pregnancy.

Pregnancy and Parent-Offspring Conflict

Human pregnancy is where life of the current generation commingles with that of the next. For thirty-eight long, hard weeks a mother cossets and nourishes her baby-to-be. How does the little fiend return the favor? By subverting her efforts at every turn. In seven of every ten pregnancies he, or especially she, succeeds in making mom retch during early pregnancy. Daughters are associated with higher levels of pregnancy sickness than sons. Demands for more food may cause dangerous rises in mom's blood pressure, or trigger gestational diabetes. Pregnancy is a turf war, the first battleground for intergenerational conflict. This conflict begins at conception, and it ends . . . well, I suppose my nine-year-old will let me know when it has ended.

Human pregnancy is often and justly called a miracle. The miracle is that it ever works. So often it does not. Between half and 95% of all conceptions end in failure, most before pregnancy is clinically recognized. (Only about 15% of clinically recognized pregnancies miscarry.) Even when pregnancy survives the first fortnight, about 40% of the time it ends in mild to serious complications, or worse. It is an awkward, ungainly, ill-designed, and painful process. A human mother undergoes extensive physiological remodeling. Over the course of pregnancy mom gets fat, and stretched all out of shape. At the end of nine uncomfortable months, a baby far too big is squeezed though an opening much too small. Tempers get lost, heads get stuck, and grudges are born. Whoever designed the process of human reproduction was clearly a misogynist, and probably failed first-year engineering. But for all its obvious flaws, human pregnancy is also the Rosetta stone for the biomedical sciences. And evolutionary biology and the theory of genetic conflict hold the key to deciphering it.

In 1993 the evolutionary theorist David Haig published a lengthy and remarkable paper, "Genetic conflicts in human pregnancy," in the *Quarterly Review of Biology*. It has lots of long words I have difficulty pronouncing and is densely packed with citations to a literature I largely do not understand. But contained within is a bold attempt to apply evolutionary thinking to the problem of human pregnancy. Why do mothers suffer from nausea and vomiting during early pregnancy? Why preeclampsia or dangerously high blood pressure? Why gestational diabetes?

Haig's work illustrates the convoluted and unpredictable paths to solving the big problems of science. There is a wide misperception that if you want to understand human pregnancy, or find a cure for cancer, or diabetes or heart disease, you tackle the problem directly. You hire mechanics and engineers to find the problem and fix it. That is pretty much the state of the art in the biomedical sciences. The labs are like multimillion-dollar repair shops with air hoses, grease guns, computer diagnostic equipment, hoists and lifts, and a complete set of socket wrenches to check under the hoods of lab rats, mice, and pigs, to diagnose and fix lupus, diabetes, cardiovascular disease, and clogged carburetors. Very bright people make inspired (and more often uninspired) guesses and occasionally get it right. Incremental progress is made. We knock another point off the mortality rate for breast cancer with a new drug therapy. This is applied science. It works, and we need more and good mechanics and engineers.

But we also need people who step back from the front lines of applied science and ask questions such as why does it work that way. The key to solving the big problems as often as not comes from unexpected directions, and by simply letting bright people tackle whatever problem they feel like tackling. More often than not the best scientists have a homing instinct for the important problems. A very good scientist, Sir Peter Medawar, once remarked that good scientists tackle the most difficult problems they know they can solve. An unlikely duo of a precocious American and a gifted but scattered Englishman tackled the problem of resolving the three-dimensional structure of a particularly vexing acid molecule called DNA, not because they knew they would found a multitrillion-dollar biotech industry and lay the groundwork for most of the Nobel Prizes in medicine to be handed out for the rest of eternity, but because their homing instinct told them it was a major problem that they could solve.

The key to solving cancer in its myriad forms lies in obtaining a basic understanding of how cells work. How does cell division get turned on and off? Learning the fundamental principles of cell biology will do more for winning this gruesome war than fighting a thousand individual battles where drug A is used against disease B with result C. We need to understand, among other things, cell ultra-structure, and how a cell communicates with itself and other cells. Most politicians, the decision makers largely responsible for research funding, are not well versed in the methods of science and the techniques of its success. They want payoffs for their research dollars and are impatient about getting results. And since there is little commercial incentive for stepping back from the front lines, why should we fund basic research? We should fund it because it provides the key insights to the big questions of science. Consider the following.

Natural-Born Cancers

The placenta that sustains and nourishes the developing baby is, in effect, a tumor that invades the lining of mother's uterus. She must engage the growing tumor swiftly, contain its advance, and hold it to a stalemate for thirty-eight weeks. A placenta is a natural-born cancer. And mother has a natural-born cure. Unraveling this complex interplay between mother and offspring *will* yield important insights into treatment for cancer. But until one adopts the perspective that mother and embryo are literally at war with one another, this obvious place to look for solutions to containing the growth of cancerous cells is not so obvious.

Moles, like cancer cells, are genetic aberrations; they can help us unravel the parallels between pregnancy and cancer. Moles arise when the maternal complement of an egg is lost and replaced by a set of paternal chromosomes. The result is that the fertilized egg carries twice the normal complement of paternal chromosomes with no maternal chromosomes, instead of equal numbers of each. The result is a seriously abnormal and potentially dangerous pregnancy. During a molar pregnancy, the placenta grows out of control until, in the second trimester, the breakdown of placental blood vessels leads to maternal bleeding. The result can be fatal if left untreated. Occasionally a molar pregnancy results in a trophoblastic malignancy, a form of cancer. Cells from the leading edge of the placenta—the trophoblast—invade uterine tissues or the bloodstream and become cancerous, again with potentially fatal consequences if untreated. Fortunately, this form of cancer is responsive to chemotherapy and thus highly treatable.

Here we see most clearly the role of male and female genes during development. Complete moles have the normal two sets of chromosomes; both male, but more than just this is needed for normal pregnancy. A typical fertilization not only combines sets of male and female chromosomes; it pits them against one another in a delicate balance. Imagine two evenly matched teams on a soccer field, say Italy and Brazil. The normal outcome of the game will be a draw, but the score conceals much that goes on during the contest. The skill of the two teams is obvious, and there is an ebb and flow to the game as the ball moves up and down the field, but because in soccer defense usually triumphs over offense, the result is still a draw. If, however, we fix the game by replacing the Italian squad with, say, the Brazilian B team, the result is far different. The game becomes a sham, particularly when both teams combine forces and spend their time scoring on an empty goal. Instead of the normal balance between offense and defense, we now observe a rout. Even with the splendid Brazilian artistry, the game quickly becomes very boring as the score reaches astronomic levels. This is pretty much what a molar pregnancy represents. During the course of a normal human pregnancy there is a balance between the potent offense of paternal genes and the stifling defense of maternal genes, and the outcome of this evolutionary game is also a draw. But in a complete mole we replace the maternal chromosomes with a set of paternal chromosomes, and the contest now becomes one-sided. There are also partial moles, with three instead of the normal two sets of chromosomes, two sets from one parent and one set from the other. Our soccer equivalent would be to match the Italian Azuri against both Brazilian teams on the same field. The game would not be a complete sham but would still be one-sided.

Sometimes it is not the entire set of chromosomes that is replaced but just a single chromosome, or even just a chunk of chromosome. Instead of replacing the entire Italian team, imagine that a single Italian player is replaced by a single Brazilian player. Instead of eleven players against eleven, it is now twelve Brazilians against ten Italians. The Brazilians gain a substantial advantage, but there is still a real match to be played. This too finds a parallel in human pregnancy. For most of the twenty-five thousand or so human genes, it would make little difference whether an individual received both genes from one parent. The system would work equally well with two copies from mom, two copies from dad, or more typically one gene from each parent. But for the roughly four dozen imprinted genes, the outcome rests critically on the parent of origin of the genes.

Imprinted Genes in Humans

Geneticists are now linking a growing number of heritable disorders in humans to imprinted genes. One such example is a pair of disorders—Prader-Willi and Angelman syndromes—that involve the same chunk of chromosome and differ only in whether the maternal or paternal region is affected. If the affected gene(s) were not imprinted, deletions on the maternal and paternal chromosomes would have an identical effect. But here they do not.

Newborns with Prader-Willi syndrome are very floppy and usually fail to suckle normally. As children they exhibit slow motor-skill and mental development, their IQ rarely rising above 80. At six to twelve months of age they develop voracious appetites and are prone to morbid obesity. Those afflicted with Prader-Willi syndrome are short in stature; they exhibit small hands, feet, and genitalia; and their puberty is usually delayed.

A different set of symptoms characterizes Angelman syndrome. Afflicted individuals suffer severe developmental delay. Their speech is either lacking or impaired; most are never able to walk; and limb movement is tremulous, often with hand flapping. They have an apparently happy demeanor, are easily excited, and are prone to inappropriate laughter. Prader-Willi and Angelman syndromes differ because of the differing effects of maternal and paternal genes on development.

There are multiple pathways to both Prader-Willi and Angelman syndrome, but they share one property: they all deactivate genes that sit on an imprinted portion of chromosome 15. This can involve deletion of that portion of the chromosome on which the genes reside, mutations to individual genes that render them inoperative, or uniparental disomy for chromosome 15 (i.e., both copies of the chromosome come from one parent). Prader-Willi syndrome involves the paternal chromosome (paternal genes nonfunctional, maternal genes functional). Angelman syndrome involves the maternal chromosome (paternal genes functional, maternal genes nonfunctional).

The strongest explanation for genomic imprinting is Haig's genetic conflict hypothesis. Here multiple paternity is key. If a father is not likely to be the father of all of mother's future offspring, then paternal genes will benefit by extracting a greater share of maternal resources than mother is prepared to give. Consider the following simple example. Assume that mother and father mate just once, and then take on new partners for all future matings. The father and his genes that reside in the embryo resulting from the current mating will not be genetically related to any of mother's future offspring, whereas mother and her genes that reside in the embryo will be equally related to all future offspring. In this case, the paternal genes should take mother for everything they can get. They have no vested interest in mom's future reproduction, and for them the future is now. If taking more food now means that the current embryo will benefit, even if it compromises mother's future survival and reproductive success, then paternal genes are for it. But the maternal genes in the embryo have a different evolutionary perspective. Mother's future offspring are their siblings and potential carriers of those same maternal genes. They cannot afford to be too selfish, as they will sacrifice future relatives.

Under the genetic conflict hypothesis, we expect paternal genes to promote the fitness of the current embryo more than maternal genes promote this. In placental mammals, an obvious route to enhancing embryo fitness is to enhance embryo growth. At the risk of oversimplifying the genetic conflict, a fatter embryo means a thinner mom, one less likely to survive and reproduce again.

Genetic Conflict and Beckwith-Wiedemann Syndrome

Beckwith-Wiedemann syndrome (BWS) is associated with an im-printed cluster of genes on chromosome 11, including the gene for insulin-like growth factor II (IGF2 gene). BWS arises either when there are two copies of the paternal gene or when the maternal genes are silenced. It is associated with fetal overgrowth and an unusually large placenta, and intriguingly those afflicted with BWS are prone to childhood tumors. It is a trait associated with paternal gene ex-pression, and is most parsimoniously interpreted as a paternal ex-ploitation of maternal genes. Insulin-like growth factor II is pro-duced by the embryo and is critical to the placental invasion of the mother's uterine lining. It is paternally expressed, the maternal copy of the gene being silent. It is regulated by an adjacent gene (H19) that is maternally expressed.

Here we have a clear illustration of the antagonism between genes inherited from mom and dad. The IGF-2 gene promotes placental invasion, in accordance with paternal interests. A maternal gene, H19, regulates the action of IGF-2 to prevent excessive invasion. H19 was in fact originally suggested to be a tumor suppressor gene, which is not far off the mark, given that the placenta, particularly its pater-nal component, is a natural-born cancer. And this may also explain why individuals with Beckwith-Wiedemann syndrome are prone to childhood cancer. Thus, in the case of BWS, dad's genes are overex-pressed. Next I shall examine the consequences of mothers being too effective at controlling placental growth in human pregnancy.

Parent-Offspring Conflict over Embryo Growth

Preeclampsia is dangerously high blood pressure during pregnancy. The medical perspective is that something is awry and needs repair. The evolutionary perspective is, why? The theory of genetic conflict provides an immediate starting point to address this question. Higher blood pressure increases blood flow through the placenta, where the fetal and maternal blood circulation aligns and is sepa-rated by a thin semipermeable membrane. More blood flow means more nutrients for the embryo. Might this be a mechanism for a baby-to-be to secure more resources for itself, perhaps more than mom wants to give? That is exactly what David Haig suggested. If he is right (and he usually is), then we would expect a placental trigger for preeclampsia, since it is the embryo that constructs the placenta. And even more specifically we would expect paternal, not maternal, genes to be responsible for this dirty deed.

Evidence is growing to support Haig's argument. First, it is the paternal complement of genes that is primarily responsible for constructing the placenta. Second, preeclampsia is especially common when the placental growth is poor and it embeds shallowly in the uterine lining. In other words, mom has defended too well against dad's placental genes. With restricted placental blood flow, and limited access to maternal fuel supplies, the embryo's prospects are poor. The solution? Crank up the blood flow. Exactly how this occurs is not yet known, but Haig has suggested where we should look.

Preeclampsia is particularly interesting because one conflict over initial placental growth may be linked to another. Indeed preeclampsia may be an embryonic retaliation for not being allowed to grow a more extensive and deeply embedded placenta in the first place. The growth of the placenta is governed by energy intake in early pregnancy. If future food supplies are likely to be low, then the embryo responds by growing a larger placenta that is better able to harvest whatever nutrients become available. One wonders if mothers at risk for preeclampsia might be able to avoid or diminish some of the symptoms by fattening up before becoming pregnant. If Haig is right, then encouraging the embryo to grow a smaller placenta by increasing the circulating food supplies may avert some of the symptoms that might otherwise become dangerous. This is conjecture, but conjecture is a useful starting point for medical research if built from a logical framework.

Haig has also suggested that preeclampsia is likely to be associated with imprinted genes, and imprinted genes appear to be involved in placental growth. The maternal gene resists the placental invasion of the uterine wall, resulting in a shallowly embedded placenta that triggers preeclampsia. The genetics of preeclampsia have proven particularly intractable to resolve, so much so that it has been called the "disease of theories." Part of the reason may be that it is associated with imprinted genes that do not obey what until recently have been the standard rules of genetics.

Imprinting and Gestational Diabetes

Haig's genetic conflict theory also provides a tidy explanation for gestational diabetes, which afflicts roughly one in twenty-five pregnancies. It is characterized by high levels of circulating glucose and insulin in the maternal bloodstream in the third trimester of pregnancy. Insulin secreted by the pancreas normally lowers blood glucose levels following a meal by increasing glucose uptake by the liver and skeletal muscles. But mothers with gestational diabetes strangely become insulin resistant, allowing blood glucose levels to soar. The reason? The fetus is working behind the scenes against its mother.

The fetus secretes its own hormones, mainly human placental lactogen, to reduce maternal sensitivity to insulin, even at high levels. The result is an elevated level of blood glucose that an embryo can draw upon for its own growth. Indeed, gestational diabetes is often associated with fetal overgrowth. And here the embryo's gain may be mother's loss, not only the short-term loss of food supplies but long-term damage to the pancreas. Mothers with gestational diabetes are more prone to developing full-blown type 1 diabetes later in life, even though gestational diabetes disappears immediately after birth. Or it might be that mothers vary in pancreatic "capacity": those with a lower capacity face a dual risk of gestational diabetes and later type 1 diabetes.

Both preeclampsia and gestational diabetes are fetal mechanisms to increase a fetus's access to food at its mother's expense. They are obvious and common manifestations of parent-offspring conflict, but not the most common. The symptoms of conflict between mother and embryo manifest themselves in an obvious way in about 70% of all pregnancies. This is what we shall turn to next.

Pregnancy Sickness and Genetic Conflict

For most mothers in most societies, food aversions and some nausea and vomiting are routine features of early pregnancy. For a small fraction, about 1%, the nausea and vomiting (NVP) are prolonged and debilitating, the form experienced by Charlotte Brontë. This is hyperemesis gravidarum, and if left untreated it can be fatal. But is there some benefit to pregnancy sickness? Might it be an adaptation rather than pathology? Several lines of evidence suggest that it is. Nausea and vomiting during pregnancy might just protect developing embryos from dietary teratogens (chemicals or pathogens that cause developmental abnormalities) during the vulnerable period of embryogenesis.

This idea is not new. In the late 1970s Ernest Hook postulated that nausea and vomiting during pregnancy yielded protection against alcohol, a potent dietary mutagen, and tobacco smoke. Margie Profet later suggested that NVP was a general protective strategy against an array of dietary toxins and pathogens, and in particular plant toxins. Unlike animals that can run away and hide from predators, plants are immobile. Some plants avail themselves of mechanical defenses including spines and thorns, but others resort to chemical subterfuge. They defend themselves chemically, and do so by dosing themselves with compounds harmless to themselves but harmful or distasteful to herbivores. Some plants produce hormonal mimics that disrupt the reproduction of herbivorous insects; others load up on chemicals such as strychnine, nicotine, and mescaline. Many plants add tannins that impart a bitter flavor. Some of these plant-borne chemicals are potentially harmful to developing embryos, especially if ingested in high doses. Not surprisingly, much of the last ten millennia of plant breeding has been designed to make food crops more palatable by lowering levels of the offending chemicals, though it seems that we still have some way to go with brussels sprouts.

Recently, Samuel Flaxman and Paul Sherman extended the discussion regarding food-borne hazards to embryos and mothers, and the possible adaptive role of pregnancy sickness. In particular they emphasize the role of spoiled and burned meats as targets of dietary aversions, and suggest that mothers whose immune systems are depressed during pregnancy need to avoid potential sources of pathogens when they are vulnerable.

There are five lines of evidence for a prophylactic role for NVP during pregnancy. First, pathogens and teratogens occur naturally in the diet, and some indeed have demonstrably harmful effects. Alcohol, caffeine, and a variety of plant toxins are known to be harmful to developing embryos. Certain bacteria, such as those responsible for listeriosis and toxoplasmosis, pose grave threats to developing embryos. While mothers are protected from many such bacteria by their own immune system, embryos must rely on maternal antibodies for protection until they are old enough to produce their own. However, the specific antibodies that protect against listeria and toxoplasmosis are too big to make the jump from mother to embryo across the placental barrier, and thus pregnant women who consume unpasteurized dairy products or uncooked meats (listeria), or who clean a kitty litter box (toxoplasmosis), may be exposing their embryos to lethal pathogens. Second, nausea and vomiting during pregnancy are associated with food aversions. Strong and pungent odors are often correlated with the presence of harmful bacteria or mutagens. Indeed, it makes perfect sense that mothers should avoid spoiled foods that may carry harmful bacteria. Third, NVP correlates neatly with the period when the developing embryo is at greatest risk: when it is developing its organs. A teratogen is a chemical that disrupts development, and embryos are most sensitive during the first trimester, when organ formation occurs. Fourth, the incidence of nausea and vomiting correlates with diet. In particular, the incidence of NVP is especially low in societies with corn- or maize-based diets but more common in societies that use meats, fish, and dairy products as staples. And fifth, nausea and vomiting correlate with pregnancy outcome. Mothers with NVP experience lower rates of miscarriage than mothers without. Perhaps this arises from the protection afforded by NVP. But perhaps not, as we shall see below.

Overall, there is certainly a strong prima facie case that pregnancy sickness serves a protective function. But might there be another explanation? There is: pregnancy sickness arises as a by-product of genetic conflict between mother and embryo. This argument is almost certainly correct, though it is not mutually exclusive of the protection hypothesis. As we have already seen, there are numerous potential conflicts between mother and embryo. This one is over whether pregnancy should continue at all. Which party decides? One might think the answer obvious: clearly it is the mother who decides. But in fact the embryo has an important, even decisive, vote in this decision. Its vote is expressed chemically and involves a hormone, perhaps *the* hormone of pregnancy: human chorionic gonadotropin (HGC).

HCG: The Hormone of Pregnancy Sickness?

HCG is a multifarious molecule in both structure and function. For the present discussion, its most important function is to maintain the pregnancy during the first trimester, and it does so by maintaining the corpus luteum. The corpus luteum is what is left over after an egg is ovulated from the ovary. It is responsible for the production of the hormone progesterone over the first weeks of pregnancy, and it is progesterone that prepares the uterine wall for implantation of an egg, and maintains it thereafter. Turn off the production of progesterone and menstruation resumes, sloughing off the uterine lining and any implanted egg. This, of course, is not in the embryo's interest. Perhaps embryos would not care about their mothers' hormone levels if moms were intent on rearing each and every embryo. But mothers are not nearly so generous. In fact, human mothers spontaneously abort between half (twenty-year-old women) and 96% (women over forty) of all conceptions, most in the first two weeks of pregnancy before they even realize they are pregnant.

Humans, other primates, horses, and perhaps guinea pigs differ from all other mammals in the production of chorionic gonadotropins. In other mammals, the corpus luteum is maintained by the mother with a hormone closely related to HCG: luteinizing hormone (LH). Among this vast majority of mammals, mothers desiring

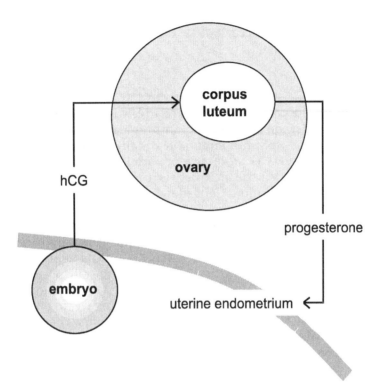

Figure 5.1. Hormonal control of pregnancy maintenance by the embryo
The embryo/placenta complex produces human chorionic gonadotropin (HCG), which stimulates the corpus luteum (the structure left over after an egg is released from the ovary) to produce progesterone. During the first trimester of pregnancy, progesterone maintains the uterine lining (endometrium), and thus is required for pregnancy maintenance. If the level of HCG is insufficient, progesterone production falls and menstruation resumes, resulting in miscarriage.

to terminate a pregnancy early on can simply shut down LH production, and that is that. But somewhere in the distant past, primate embryos picked up a neat trick. They stole control of pregnancy from their mother. Or more precisely, they replaced LH production with HCG production. LH is in fact the parent to HCG. These are two members of a family of hormones known as dimeric glycoproteins. What that means is that the hormone is made up of two chains, alpha and beta. The alpha chains in this hormone family are pretty much interchangeable. It is the beta chain that differs in structure and determines the biological function. HCG is very similar to LH but for the addition of a tail to the beta chain. The presence of this tail and a few other minor structural alterations are critically important to the story. If not for the tail, HCG would be toxic to mother in the quantities produced. But let us return to the origin of this biochemical coup d'état.

Evolution of Chorionic Gonadotropins in Primates

Chorionic gonadotropins (CG) are a relatively late arrival in the primates. They first appeared in the anthropoid primates, after the monkeys and apes split from our more distant cousins, the tarsiers, lemurs, and lorises. CG arose as a duplication of the original LH gene, and it preceded the split between the Old World and New World monkeys. Two more CG genes were added in the common ancestor of the langurs, macaques, mandrills, humans, and apes. A fourth gene shows up in the ancestor of humans and gorillas, and two more after that. Thus gorillas have four CG genes, humans six. At the same time, the original LH molecule was being modified with the addition of side branches that serve to slow the rate of clearance of CGs from mother's bloodstream. LH is a relatively short-lived molecule in maternal blood, with half of it being cleared in thirty minutes; with HCG that same half-life is twelve hours. That molecular change makes HCG a more potent weapon for an embryo, and more dangerous to mother.

The evolution of CGs is closely associated with placental structure. Most mammals possess an epitheliochorial placenta, the equivalent of a placenta with a childproof cap. Humans, monkeys, and apes possess a hemochorial or easy-access placenta. The difference is a double layer of tissue. The human/ape placenta is in direct contact with maternal blood. Gases, nutrients, and hormones pass relatively easily between mother and embryo. Not so with a childproof placenta, in which maternal blood is separated from fetal blood by two extra layers of tissue that impede the flow of large molecules—and in particular hormones—from the placenta to the mother. In practical terms, an epitheliochorial placenta means that an embryo cannot easily manipulate its mother with placental hormones. A human or chimp or gorilla embryo can. Are they guilty of this chemical sin? Well, they have motive (avert miscarriage), means (genes to manipulate mother), *and* opportunity (access to maternal blood).

Chorionic Gonadotropins and Miscarriage

Miscarriage is common during mammalian pregnancy. Perhaps a mother has detected an embryo with a genetic defect, or nutritional conditions have taken a turn for the worse. Mother's best option is to cut her losses and begin anew at some later date, withdrawing maternal support for the current embryo. The mechanism is simple: shut down production of LH. But this simple mechanism is open to embryonic subversion, as any gene that a mother has, an embryo has too. Thus if mom has genes for manufacturing LH, so too does the embryo. What is to stop an embryo from turning on these genes early to avert what for mom is an adaptive miscarriage? Nothing, really.

Somewhere in our evolutionary history, embryos picked up this neat trick. They set their own genes for LH production to the "on" position while still an embryo. By adding to the maternal supply of LH, embryos that would have perished in a miscarriage now survived. This neat trick carried a huge evolutionary advantage and would have spread rapidly. But mothers were not disinterested observers in this evolutionary game. They too were subject to strong selection. Mothers that could regain control of pregnancy also gained an advantage. Thus, if embryos produced more LH, mothers resisted, either by reducing their own production or by reducing their sensitivity to the molecule. Hormones are chemical messages, and they interact with receptors that are found on and in cells. The communication system can be regulated either by changing the level of hormone or by changing the number of receptors. Fewer receptors mean a lowered sensitivity to a hormone.

By reducing her sensitivity to LH, mother potentially wrested control of pregnancy away from the embryo. But such an advantage was likely to be temporary, as embryos were likely to retaliate by producing ever more LH. There was one potential stumbling block for the embryo. LH is normally secreted at low levels, and it is a thyroid stimulator. Producing high levels of LH would overstimulate the thyroid, resulting in thyroid disease, and this is normally incompatible with pregnancy—causing, for example, miscarriage, premature birth, or retarded fetal growth. Thus, there were limits to what an embryo could do. If LH levels rose too far, an embryo risked poisoning mom and, in doing so, itself too.

Natural selection, though, is ever resourceful. Somewhere in the distant evolutionary past of primates, the original form of LH mutated to a new form, one in which a tail was added to the molecule. This and a small bit of tinkering with the three-dimensional structure of the old hormone resulted in a new form that differed from the original LH molecule in one key respect. It lost most of its ability to stimulate the thyroid. This is the point where the ancestral LH molecule became a chorionic gonadotropin. Such alterations are the bread and butter of molecular evolution. The genes for chorionic gonadotropins arose from duplications of the original gene for LH,

and the key genes responsible for the part of the molecule that determines biological activity sit together in a cluster on human chromosome 19; there is one gene for LH and half a dozen for HCG. Now
that HCG has been rendered nontoxic to mother by altering the
structure of the original LH molecule, the embryo can produce it in
much greater quantities than LH. And it does!

Primates had entered into an evolutionary arms race, a period of
retaliation and escalation. In such a race, each time the stakes are
raised by one party, the other counters that move and then raises
the stakes higher still. One can easily imagine a series of steps: The
embryo raises its hormone production, and mother responds, by reducing her sensitivity to it. The embryo in effect jumps higher, so
mother responds by raising the bar. Where does such an arms race
end? When it becomes too costly for one or both parties to continue.
And that cost may just be pregnancy sickness. During the Cold War,
the Soviet Union and the United States engaged in an arms race
that resulted in a massive arms buildup, ending evidently when the
Soviet economy could no longer sustain the heavy cost. Human
mothers and embryos also appear to have reached the end point
where further escalation is unsustainable: hyperemesis gravidarum.

Most pregnancy sickness is mild and has no permanent effect on
either the mother or the embryo. But such is not the case with hyperemesis gravidarum. It is potentially fatal if left untreated, and it is
only with the advent of treatment via intravenous fluids that
women have stopped dying from it. The system has rolled to a stop
where most human mothers sit just beneath the threshold of heavy
costs; a few cross the line. If the arms race had continued, dangerous
levels of pregnancy sickness would afflict many more women. As
things now stand, it is relatively rare, occurring in less than 1% of
all pregnancies.

David Haig suggests that such a chemical arms race is exactly what
has happened in humans, and as he so eloquently puts it, "raised
voices are almost always signs of conflict." Human embryos are
screaming with deafening levels of HCG, so much so that the excess
even spills into a mother's urine, where it becomes an early, reliable
cue of pregnancy. The result of all this screaming? Pregnancy sickness. Though its thyroid-stimulating activity is greatly diminished,
HCG is produced at such high levels that it still affects the thyroid,
and this may be the key link to understanding pregnancy sickness.
Moreover, there is not just one form of HCG but several, and some
are more active thyroid stimulators than others. The form of HCG
produced by hydatidiform moles and Down's embryos is especially
potent, and both molar and Down's pregnancies are associated with
high levels of hyperemesis gravidarum.

HCG has long been suspected as the key hormonal trigger for pregnancy sickness, but the evidence has been mixed. A consensus, however, finally appears to be emerging that HCG is important and perhaps decisive. The most obvious link is the simple fact that the production of HCG in early pregnancy is correlated very precisely with the timing of pregnancy sickness. As soon as an embryo implants, the developing placenta oozes HCG, releasing it into the mother's bloodstream. Blood levels of HCG rise rapidly until the eighth week of pregnancy and then begin to fall, steeply after week ten and remaining low from week fourteen until birth.

HCG is also a trigger for miscarriage. Embryos that produce low levels are aborted at high frequencies. Embryos that produce high levels are rarely lost. And here is where the hypothesis that pregnancy sickness evolved to protect the embryo against dangerous foods stumbles. HCG production is a measure of the genetic quality of an embryo. When one examines the embryos that are spontaneously aborted during the first trimester, about 70% exhibit chromosomal abnormalities, and these are present from conception. And this number is surely too low, as the methods used to detect genetic defects are too coarse to pick up other gene mutations that are equally lethal. For all we know, the true rate of genetic defect may be closer to 100% than 70%. Thus, it is not pathogens and teratogens in the diet that cause early spontaneous abortion (and are associated with low HCG production) but genetic defect. And the link between normal pregnancy sickness and lower rates of miscarriage is not the result of mother avoiding harmful foods but rather a result of the fact that the low-quality embryos that cannot produce enough HCG to make mother sick are also those doomed to be spontaneously aborted.

But whether there is a direct path from HCG to pregnancy sickness is still debated. Some workers have failed to find evidence of a link between elevated levels of HCG and normal nausea and vomiting during pregnancy, or its more extreme form, hyperemesis gravidarum, though a greater number of studies have found such a link. If HCG was involved, one might think you could measure its level, the level of pregnancy sickness, and that would be the end of the debate. If only it were so simple.

Part of the problem is it is more than just the quantity of HCG produced, but the type. The forms produced vary across individual mothers and across time during the same pregnancy. There is also the problem of accuracy of measurement: HCG levels increase and then decrease very rapidly. Unless the date of conception and hence age of gestation is known precisely, the data gathered will be of extremely poor quality. Then there are the confounding factors. Mothers who smoke are less prone to pregnancy sickness. It matters whether the corpus luteum rests on the left or right side. Underweight mothers experience less pregnancy sickness, older mothers more. All of these cloud the link between HCG and pregnancy sickness.

The circumstantial evidence surrounding HCG as the trigger for pregnancy sickness, however, is strong. First, female embryos produce more HCG and are also associated with higher levels of hyperemesis gravidarum. This link between fetal sex and pregnancy sickness is particularly intriguing, as more than two thousand years ago, the Greek philosopher Hippocrates wrote "a mother carrying a female baby has a pale face, whereas if she is carrying a male baby she has a healthy tone to her skin." It took a research team at Stockholm's Karolinska Institute, reviewing all Swedish births between 1987 and 1995 (over a million in all), to confirm Hippocrates' observation. They found a slight male bias in live births—51.4 males to 48.6 females, but a strong female bias in mothers admitted to hospital with hyperemesis gravidarum—44.3 males to 55.7 females. Severe pregnancy sickness was much more likely when a mother was carrying a girl. This effect again points directly toward HCG as the proximate cause of morning sickness, as female fetuses produce more than males. Similarly, twin gestations produce both high levels of HCG (two to ten times that in singleton gestations) and high levels of hyperemesis gravidarum. Down's embryos also produce abnormally high levels of HCG and trigger high levels of pregnancy sickness. Indeed, the fact that Down's embryos produce a high level of HCG is likely the very reason that they are so common at birth relative to other chromosomal defects. Hydatidiform moles, which secrete high levels of HCG, produce high levels of pregnancy sickness. The circumstantial evidence for a link between HCG and pregnancy sickness is compelling. Indeed, if this were evidence in a Texas murder trial, HCG would surely be on death row.

But what of the embryo/maternal protection hypothesis as an explanation for pregnancy sickness? The links between diet and the food aversions and cravings that occur during human pregnancy are too pronounced to be mere accidents. There are two separate issues here. First, do human mothers make adaptive changes to diet choice during pregnancy? And second, does pregnancy sickness per se provide dietary protection for mothers and embryos? If allowed to cast a vote here, I would register in the affirmative for the first, though even here the evidence is circumstantial. But as Thoreau stated, some circumstantial evidence—as when you find a trout in the milk—is very powerful. Moreover, it would be astonishing if natural selection has *not* had a hand in shaping maternal diet choices during pregnancy. But as for whether NVP per se is designed to protect mothers and embryos from dietary mutagens and pathogens, we have yet to find the trout in the milk. The key link between the embryo protection hypothesis and miscarriage is spurious. But it is entirely possible that natural selection has been at work here, superimposing adaptive dietary changes on an underlying genetic conflict. One possibility is that pregnancy sickness exaggerates preexisting and adaptive maternal changes of diet. If that is indeed true, then pregnancy sickness may serve as a useful probe for unraveling the nature of the underlying adaptation.

But the link to HCG as a trigger for pregnancy sickness grows ever stronger and more compelling. It appears that pregnancy sickness may not arise just from maternal/embryo conflict but chiefly from the paternal genes in the embryo. There is very strong evidence to suggest that HCG is produced by imprinted genes. The strongest evidence comes from the study of molar pregnancy.

Moles occur in 1 of every 1,500 to 2,000 pregnancies, and there are two types: complete and partial. A complete mole arises when a sperm fertilizes an empty egg (or the egg nucleus is nonfunctional). Here something curious happens. The chromosomal complement in the sperm duplicates itself. Thus, complete moles contain two complete sets of paternal (dad's) chromosomes. But there are also partial moles that contain not two sets of chromosomes, as is usual in a fertilized egg (one set from mom, another from dad), but three sets. That is, they are triploid. They have one set of chromosomes from one parent but two sets of chromosomes from the other parent. Thus you can have partial moles with two sets of genes from dad and one from mom, and partial moles with two sets of genes from mom and one from dad.

These aberrant genotypes in moles tell us about the control of HCG production. Complete moles produce the highest levels of HCG; partial moles with two paternal genomes produce intermediate levels; and partial moles with two maternal genomes the lowest levels. The quantity of paternal genes appears to regulate the quantity and perhaps the type of HCG produced. This pattern is what we would expect from an imprinted gene.

Is HCG controlled by imprinted genes? It seems so. The genes for producing the beta chain of HCG, the important bit, reside in a cluster on human chromosome 19. The same chromosomal region in mice is known to be imprinted, though that is not yet confirmed in humans. If indeed HCG proves to be imprinted in humans, it suggests that the paternal genes are more interested in averting miscarriage than maternal genes. This is consistent with Haig's genetic

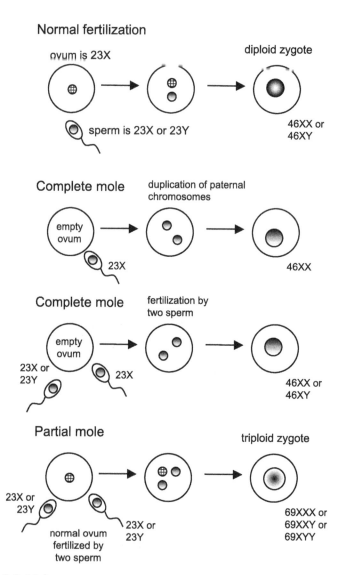

Figure 5.2. Molar pregnancy

The traditional view of molar pregnancy is an aberration that occurs at fertilization. Normal fertilization combines the chromosomes of a sperm and egg (ovum), resulting in a zygote with 46 chromosomes, either 46XX (female) or 46XY (male). Or perhaps an empty ovum (no chromosomes from mother) is fertilized by a sperm, followed by a duplication of the paternal chromosomes, which combine to form a zygote with 46 chromosomes (though evidence for this mechanism is scant). These are always 46XX, because zygotes with two Y (male sex) chromosomes are not viable. This is one path to a Complete Mole. A second path involves the fertilization of an empty ovum, by two sperm, at least one of which has a 23X chromosomal complement, as at least one X chromosome is required for a viable zygote. Occasionally two sperm fertilize a normal ovum resulting in a triploid zygote with a double paternal genetic complement. This is a partial mole. Molar pregnancies are significant because they illustrate the normal balance between maternal and paternal genes. When this is disrupted, a seriously abnormal pregnancy arises. A new view of molar pregnancy suggest that they may arise as a postfertilization byproduct of abnormal twinning (see conflict theory of imprinting, as genes from mom are more interested in mom's future reproduction.

Thus the link between HCG and Down syndrome becomes ever more compelling. It is the most common chromosomal birth defect, something I shall have more to say about in chapter 7, and arises when an error in cell division results in an extra twenty-first chromosome, hence the technical name trisomy 21. Down's embryos secrete abnormally high levels of HCG—they just ooze the stuff—and this indeed forms the basis for early detection by first-trimester serum screening. Is it just coincidence that Down's embryos are particularly good at bypassing mom's built-in screen for birth defects? Almost certainly not. And this ability might also have something to do with the dual role of HCG in maintaining the corpus luteum and suppressing the maternal immune system, two prerequisites for pregnancy maintenance. But most intriguingly, is there a link between maternal age and genomic imprinting? If maternal genes are interested in protecting mother's future reproductive success, more so than paternal genes, what happens as mother approaches menopause, after which there is no reproductive future to protect?

Chapter 6

Selfishness Unconstrained

LOKI THE SHAPE-SHIFTER is the most intriguing character of Norse mythology. He is the trickster who keeps the company of gods, whom he sometimes angers and sometimes pleases. Loki persuaded the gods to accept the challenge of a solitary mason whose task was to rebuild the stone wall surrounding Asgard, the home of the gods. Should he succeed, the mason demanded the hand of Freyja, fairest of all the goddesses, in marriage. Six months was the time allotted to the seemingly impossible task, for Asgard was vast in expanse. The gods were convinced by Loki that they could not lose, and with their wall rebuilt (a task they would finish when the mason failed), they would again be safe from their enemies, the rock and frost giants. But the mason, a rock giant in disguise, far exceeded expectations, and with the six-month deadline looming, the wall was nearly complete. The gods grew concerned and angry, and demanded that Loki do something to save Freyja from shame.

Loki succeeded. He used trickery to deprive the mason of his workhorse, and the wall could not be finished. Loki the troublemaker had tempted fate and won, but just barely. His nature was to take pleasure in the misfortune of others, and within Norse mythology he is transformed gradually from trickster to demon. Loki's treachery led to the death of Balder, most beloved of all the gods and son of Odin, most powerful of all gods. When Odin arranged to rescue Balder from the realm of the dead, Loki thwarted the effort. He had then gone too far and knew it. He fled Asgard to avoid the wrath of the gods. And though he could run, there was nowhere to hide from Odin the all-seeing, even as a shape-shifter disguised as a salmon. Loki was hunted down, captured, and ushered to a dank, dimly lit cave. The gods exacted revenge by tethering Loki beneath a snake dripping venom.

Loki was selfish and greedy, and that he was tolerated by the gods, a capricious and vengeful lot, is all the more surprising, since he was not one of them. He did not share their bloodline, as he was the son of giants and foster brother to Odin. Loki therefore needed his guile and sharp wit to overcome their natural enmity toward him.

Why do we expect better treatment from relatives? The short answer is shared genetic interests. Taking too much from a brother or sister can do more harm than good because it deprives relatives who share identical genes of needed resources. The behavior that emerges is the product of an enlightened, genetic self-interest. No such check on selfish behavior exists for nonrelatives. Among birds, fish, and insects are species that make their living by stealing parental care from nonrelatives, most often from conspecifics but sometimes, most spectacularly, from other species entirely. These are brood parasites.

Brood parasitism is of great interest to evolutionary biologists because it can be used as a probe to study the normal limits of selfishness. Without relatedness to constrain selfish behavior we descend into a Hobbesian war of all against all. The conflicts between biological parents and their children are both common and obvious: children plead and squabble among themselves; parents arbitrate and discipline. But how much worse could it get? The study of brood parasitism provides the answer. And best known among the brood parasites are the brood parasitic birds.

Brood Parasitic Birds

About 100 out of the 9,000 or so bird species are obligate brood parasites. That is, they make their living by exploiting the parenting instincts of the other 8,900. The brood parasites build no nests and rear no families. Instead they lay their eggs in the nests of other birds and leave the brood rearing to the foster parents. Like Loki, brood parasites rely on trickery and deception to survive. Avian brood parasites are useful models for the evolution of selfishness, as the foreign nestling is unrelated to its brood mates, and hence unconstrained by the tempering influence of kin selection.

Old World Cuckoos

The common cuckoo of Eurasia surely sets the standard for extreme selfishness. Cuckoos are the best known of the Old World brood parasites and are the product of a long coevolutionary arms race with their hosts. Common cuckoos, which range widely across Eurasia, parasitize many species, but individual cuckoos specialize on a single host. Their trump card is the mimetic egg: their eggs appear to the human eye almost indistinguishable from those of the host. The hosts also seem to have difficulty telling the difference, as cuckoo eggs are readily accepted within the clutch even though the cuckoo chick will execute the hosts' own progeny. Immediately after hatching, a cuckoo chick ejects all other eggs and nestlings, leaving it as the sole occupant of the nest and exclusive recipient of its foster parents' largesse. Even its call is designed to mimic a full brood of nestlings, to induce parents to work hard on behalf of the interloper.

Recent work on spotted cuckoos in Spain suggests a further chilling dimension to host-parasite relations: an avian mafia. Hosts that reject cuckoo eggs may be punished by having their nests destroyed, as though the cuckoos are saying, "If you don't rear my chick, I won't let you rear yours." When cuckoo eggs were experimentally removed from host nests, the rate of nest destruction rose sharply over the destruction rate where the cuckoo eggs were untouched. Many including myself are skeptical, but should this work stand, it adds a remarkable twist to the story of foster parenting in birds.

. . . and New World Cowbirds

The New World counterpart of the Old World cuckoo is the cowbird. It lacks a cuckoo's flair but is equally dangerous. Brown-headed cowbirds are nondescript residents of pastures and fields, similar in size and appearance to a typical blackbird—not surprising, since they are one of the hundred or so species of New World blackbirds. There is little outward sign of their sinister parenting, yet they too are obligate brood parasites. They too inflict their offspring on foster parents but, unlike common cuckoos, make no attempt to conceal their presence through egg mimicry. A female wood thrush may return to her nest to find a brown speckled egg sitting conspicuously amid her clutch of pale blue eggs. Worse still, one of her own eggs may be missing. Cowbirds simply add their own egg to the clutch of the foster parent and, to make room, often remove a host egg.

Speed is part of the cowbird's strategy. Hatching at the same time or preferably before the host nestlings is key to their success. Size is important, since the nestling with its neck stretched highest is most likely to be fed first. The best way to ensure this favorable outcome is to hatch first after a short incubation period—only ten to eleven days, which is at the minimum for all birds. With a head start in development, a cowbird nestling can hold its own in the nests of larger hosts such as red-winged blackbirds and quickly dominate nestlings of smaller hosts such as yellow warblers, often with lethal consequences.

Cowbird Mafia?

Cowbirds have been honed by natural selection for ruthless efficiency. Adults survey the nesting activities of their hosts to ensure that eggs are laid at appropriate times. But if the potential host has already laid a complete clutch and commenced incubation, a cowbird loses its opportunity to add one of its own offspring. The solution? Some cowbirds destroy unparasitized host clutches, inducing the host to lay a new clutch, one that can be parasitized. It was not until recently that ornithologists realized that cowbirds are not only brood parasites but also predators of eggs and nestlings. Their interest as a predator, though, is not gastronomic: they are being entrepreneurial, creating new clients for their product. Cowbirds are much like a gang of thieves who are also in the business of selling burglar alarms. The hosts are clear losers in this evolutionary game.

Cuckoo Catfish

Brood parasitism reaches a sinister extreme not in birds but in fish. Cichlids are one of several groups of fish in which some species care for their offspring by mouth brooding. The eggs and fry are carried in the mouth of the parent, where they are protected from pathogens, parasites, and predators. Cuckoo catfish, whose eggs are picked up by the cichlid, sometimes parasitize mouth-brooding cichlids living in Lake Tanganyika in the African Rift Valley. As in cowbirds, the parasite eggs develop more rapidly than the host eggs, and upon hatching then eat the host eggs. Who said cheaters never prosper?

The Origins of Brood Parasitism

Brood parasitism represents the ultimate in avian day care. It starts before birth, is perpetual, and, best of all, is cost free. Once the egg is left behind, the brood parasite's parental responsibilities end, and the foster parent's begin. But there is truth in the old maxim you get what you pay for. Brood parasite nestlings do not fare as well as the host nestlings, for a variety of reasons. But the beauty of a brood parasitism strategy is that they do not have to. Freed from the burden of having to incubate eggs and care for nestlings, brood parasites can produce eggs in industrial quantities. Most songbirds produce clutches ranging between three and eight eggs. A single cowbird female is capable of laying more than forty eggs during a breeding season, though recent work suggests that the actual number laid is far lower. If only a fraction of these survive, the cowbird comes out ahead. But how did the habit of stealing parental care ever get started? It almost certainly began with the parasitism of conspecifics, or as a strategy of facultative brood parasitism.

Redheads are prairie-nesting ducks that are not obligate brood parasites, as they build their own nests and care for their own offspring. But they sometimes dish a portion of the parenting duties off to other ducks, and closely related canvasbacks in particular. Redheads do not resort to subterfuge to succeed but instead use thuggery. They simply shove the canvasback hen aside and lay their egg. The rest is left up to the canvasback, who incubates the eggs and cares for the ducklings. Interspecific brood parasitism in ducks is not quite as sinister as brood parasitism in songbirds. Ducklings, being precocial and self-feeding, are low maintenance offspring. After hatching her eggs, the main job of the hen is to stand guard over her brood, and it is almost as easy to care for eight as it is for seven. The chief cost to the host occurs before hatching, as an extra parasite egg may cause overcrowding of the clutch. Larger clutches are incubated less efficiently, and fewer of the host's own offspring may survive.

Egg dumping is a common practice within species in many birds: a female will often sneak an egg into another's clutch. It is particularly common in ducks that leave the nest unguarded during egg laying. If there are five eggs present, who will notice if I add just one more? This logic of dilution sometimes runs amok as larger clutches become more attractive targets for brood parasites. If thirty eggs are present, who will notice a thirty-first? Some of these "dump nests" fill up with many more eggs than can ever be incubated successfully and are abandoned by the original parent. Nearly one hundred eggs were recovered from one fulvous tree duck nest, about ten times the normal clutch size. Needless to say, none survived.

Slipping an egg or two into the nest of another member of the same species is an obvious starting point for the evolution of brood parasitism. Over time, we would expect adaptations to hone the parasitic abilities of the species, but also counteradaptations for resistance. Parents robbed of parental care that would have been spent on their own offspring would quickly find themselves at an evolutionary disadvantage compared with the thieves, favoring escalation in this ongoing war. Taken to the extreme, obligate brood parasitism is a specialist strategy. In theory, some members of a species could exist as obligate brood parasites, but only as long as others remained hosts and raise the babies. But such a system seems inherently unstable and unlikely to endure. More likely, brood parasites begin to probe beyond the boundaries of their own species. And the obvious initial targets are closely related species with similar ecologies.

Adopting Runaways?

Brood parasitism is forced adoption. But sometimes it is not parents that volunteer their progeny for adoption, but runaway children that volunteer themselves. Asynchronous hatching in birds ensures that last-hatched "runts" will suffer reduced growth and elevated mortality compared with their older sibs. In many nests it quickly becomes clear that the runt is doomed to an early demise. These bleak circumstances sometimes drive the runt to quite literally run away from home, seeking refuge in another family. But other families are alert to these trespassers and set up a perimeter defense around the nesting territory, resisting with deadly force. The slim hope of the wandering chick is to dart into another brood when a parent's guard is down. Surprisingly, many species do not recognize their own offspring, at least when they are young, and rather use a simple behavioral rule based on location: care for young birds inside your territory, but drive away all those attempting to gain entry. Wandering chicks pursue a high-risk strategy with a low rate of success, but with death virtually certain at home, even a slim hope is better than none. A few do succeed, and in some cases their prospects improve dramatically, as when they go from being the runt back home to the largest within the new brood. With this change in status they acquire all the benefits of avian primogeniture.

Forced Adoption of Nonkin

Adoption is not uncommon in nature, a topic reviewed at length in Sarah Hrdy's masterful *Mother Nature*. Among many mammals, relatives may assist in the care and protection of each other's offspring. This behavior can be explained by the straightforward calculus of kin selection: assisting the progeny of close blood relatives can yield evolutionary dividends, since you all share copies of at least some genes. Adopting a relative's offspring can also allow an individual to hone his or her parenting skills on the children of others. The adoption of nonrelatives is, however, rare. The reason is straightforward: the children of others carry no genes of your own. To borrow Richard Dawkins's metaphor, children are the vehicles that carry a parent's genetic immortality. Foster kids carry no cargo but are still expensive. Natural selection results in discriminating parents who invest carefully, and prune ruthlessly, to maximize returns in a war of all against all. Any advantage, any edge, will be found by an ever creative natural selection. Any cost, any encumbrance, any handicap will be as ruthlessly weeded out. A brood parasite succeeds by robbing from another parent's reproductive future.

Parental care is precious, to be meted out carefully (or so I tell my sons Spud and Lumpy) and jealously guarded against unauthorized use. This provides a powerful incentive for brood parasites to cheat the system, and they do so by deception, guile, speed, and even brute force. The potential host, the mark in this evolutionary sting, counters with vigilance. The ongoing struggles of brood parasite and host are textbook examples of the process of coevolution, an evolutionary game of tennis in which an adaptation in the parasite is returned with a counteradaptation from the host, a volley back from the parasite with a counter-counteradaptation, a drop shot from the host, and so on. The point continues until either the host or parasite goes extinct, or the host perfects a defensive system. Species that routinely eject cowbird eggs tend not to be parasitized by cowbirds, suggesting that good defense ultimately triumphs over good offense in the game of brood parasitism.

Potential hosts resist either by preventing brood parasitism from occurring or by removing or discriminating against the parasite offspring. Hosts drive cowbird females away from nests during the vulnerable egg-laying period. The life of a cowbird can be dangerous, as larger hosts may injure or kill the intruders. Colonial nesting blackbirds enjoy the benefits of group vigilance and defense, and are rarely parasitized. Individuals nesting solitarily or at the edges of colonies are subject to much higher rates of parasitism.

If the brood parasite breaches the defensive perimeter of the host, countermeasures are deployed. Larger birds such as grackles and orioles ruthlessly eject parasite eggs from their nests, the ghost of selection operating on brood parasitism past. Smaller hosts such as warblers and vireos face greater difficulty, as their bills may be too small to pick up a cowbird egg. Some resort to puncture ejection: the host spears the cowbird egg with its bill to remove it from the nest. Such a maneuver is dangerous, as a misplaced bill peck or splattered egg yolk can damage the host's own eggs. And the cowbird makes puncture ejection more difficult by building eggs with unusually thick shells. Some small hosts are unable to remove cowbird eggs. Forced into one of a series of less desirable options, parent vireos and warblers face a cruel bind—either abandon their own clutch when victimized by a cowbird and forfeit an opportunity to breed, or rear the brood, cowbird and all. For some birds with short lives, options are desperately limited: they may not survive to breed again.

Some other hosts—such as red-winged blackbirds—could easily remove a cowbird egg but do not. Why is this so? One explanation is evolutionary lag. New hosts stand defenseless against fiendish cowbirds. To humans the solution is obvious, but to a songbird it is not. Recognizing and removing foreign eggs is a complex behavior, and natural selection, though powerful, does not look ahead and cannot anticipate what might happen. It is about what has happened already.

Avian ecologists can study the early stages of the coevolutionary process when a brood parasite meets a novice host. We are presently witness to such an event in the form of a large-scale avian invasion of the Caribbean by South American shiny cowbirds. They have expanded their range rapidly over the twentieth century, reaching Grenada in 1901, Saint Vincent in 1924, Martinique in 1948, Puerto Rico in 1955, and Cuba in 1982. By the late 1980s they had even established a beachhead in south Florida. This rapid advance has brought cowbirds into disastrous contact with songbirds unprepared for their brood parasitic habit. Species such as the Martinique oriole and the yellow-shouldered blackbird of Puerto Rico have been driven to the brink of extinction by recent cowbird parasitism. Bronzed cowbirds are now expanding their range into Texas, and brown-headed cowbirds were unknown along the west coast of the United States and Canada until the twentieth century. These recent invaders enjoy an abundance of unprepared hosts and, at least temporarily, hold the upper hand in this evolutionary game.

The brood parasites succeed by forcing adoption on naive hosts. Such selfish behavior is not hard to explain. A greater challenge is the apparently unselfish, voluntary adoption of nonrelatives. Here the third prong of parental optimism—facilitation benefits—helps to explain unexpected behavior.

Voluntary Adoption of Nonkin

Under facilitation, the presence of surplus offspring yields benefits to the host parent or offspring or both. Darters are small fish native to streams in central and eastern North America (the group includes the endangered and infamous Tennessee snail darter). Males build nests and court females, and there is one trait that females find particularly alluring: the presence of eggs in a male's nest. For a female, egg number is a cue to the quality of paternal care. With more eggs a male is more likely to work harder to ensure brood success. It is his job to fan the eggs, clear the nest of debris, keep fungal infections at bay, and fend off predators. Given that females use this cue to decide where to lay their eggs, the males have devised a remarkable tactic: they steal eggs from other males' nests. The presence of the surplus eggs makes the thief's nest even more attractive to potential mates. But egg stealing is not restricted to mating fish. It is a tactic used to create a workforce of slaves.

A dramatic exception to the general rule that adoption of nonrelatives in nature is rare occurs in the social insects. Here, adoption is not only routine in some species; it is compulsory. It occurs in slave-making ants that rob the pupae and larvae from colonies of other, related ant species to assimilate these progeny into their own collective. These ants could have been the biological models for the Borg of *Star Trek* infamy. Scouts from the slave-makers' colony strike out to locate potential targets for their armies. Once a likely target is found, they return to their own colony, leaving a chemical trail behind for the raiding armies to follow. Mobilized to action by the scouts' return to the colony, worker ants of the slave-making species march in columns to their target. At the colony to be attacked, some ants who detect the scent of a forthcoming invasion, resigned to the fact that resistance is indeed futile, may attempt to flee with their pupae or larvae held carefully in their jaws. Other workers may stay behind to defend their colony but do so at their own peril. There is an insect chivalry in these battles. The raiders kill the slave workers only if they offer a hostile defense; otherwise they are ignored in Borg-like fashion. The pupae and larvae left in the colony when the raiders arrive are captured and returned to the slave-makers' colony. There the "adopted" offspring complete development and are then absorbed into the workforce. Tasks that are normally performed by ant workers—foraging, nest building, and brood rearing—are left to the slaves. Here adoption serves an ergonomic purpose, to recruit a cheap workforce to the ant collective. This represents the ultimate in facilitation benefits, which I shall explore in greater detail in chapter 11.

The Lesson of Brood Parasitism

Brood parasitism knows no parallel in humans. The presence of an extra baby is one of those things that human parents tend to notice. But there is a lesson we can learn from brood parasites, and it is this: parent-offspring relations could be so much worse. Selfishness within the human family takes a benign form compared with that in brood parasites. The shared genetic heritage of family members holds our base instincts in check. Yes, children are demanding of parents, and siblings squabble. But human infants do not deliberately smother other infants. Older brothers or sisters do not roll younger children out of their cribs as do cuckoo nestlings, nor do they dine on their younger companions as do brood parasitic catfish. And even though it may not seem like it, most children do not have insatiable appetites, as do nestling cowbirds. And other parents do not murder children so that they can substitute their own. . . . Or do they? I shall return to this question in chapter 11.

Chapter 7

Screening for Offspring Quality

*Nor was it in the power of the father to dispose of the child as
he thought fit; he was obliged to carry it before certain triers
at a place called Lesche; these were some of the elders of the
tribe to which the child belonged; their business it was carefully to view the infant, and, if they found it stout and well
made, they gave order for its rearing, and allotted to it one of
the nine thousand shares of land above mentioned for its
maintenance, but, if they found it puny and ill-shaped, ordered it to be taken to what was called the Apothetae, a sort
of chasm under Taygetus; as thinking it neither for the good
of the child itself, nor for the public interest, that it should be
brought up, if it did not, from the very outset, appear made to
be healthy and vigorous.*

—Plutarch, *Lycurgus*

The Logic of Progeny Choice

One's fate was determined early in ancient Sparta. A council of elders
screened each newborn for defects. Fathers were ordered to rear babies found stout and sturdy, and the infant was assigned one of the
nine thousand shares of communal property. But those found wanting were cast into a ravine—an *apothetae*—at the foot of Mount
Taygetus to perish. Spartan decision making was a conscious decision to protect the public good from the perils of defective progeny.
One wonders whether Kafka could have produced a more nightmarish tandem: committees and infanticide.

More privately, human mothers practice an unconscious version
of the same progeny screening, but it takes place behind the scenes
in the first days and weeks of pregnancy, often without a mother
ever having been aware that she was pregnant if only for a brief time.
Depending on a mother's age, somewhere between one half and
96% of human conceptions end in early miscarriage. Most of these
are "occult" abortions and underline the fact that a profligate level
of pregnancy wastage is a routine feature of human reproduction.
The obvious question is, why? The simple answer is quality control.

Spontaneous abortion is the outcome of a system of maternal screening for genetic defects that are surprisingly common in humans—perhaps as many as one half of all conceptions bear chromosomal defects. Pregnancy is enormously expensive for human mothers, and natural selection has built a system that ensures that scarce resources are not squandered on low-quality progeny. Compared with ocean sunfish or orchids, which may broadcast hundreds of millions or even billions of tiny progeny into the currents of water and air, human mothers invest massive quantities of resources in each and every surviving offspring. With so much at stake, mothers must make prudent decisions, and thus hold a very conservative view about which progeny to carry to term. In this respect humans are no different from snails or sharks or the vast array of other organisms whose parents routinely create many more incipient progeny than are ever reared to independence. If the surplus progeny do not succumb to intrinsic defect or extrinsic hazard, parents then neglect, abandon, or even kill the supernumeraries. Doing so consolidates parental investment in a smaller number of higher-quality progeny, a process known as progeny choice.

Sequential versus Simultaneous Progeny Choice

Progeny choice can be either sequential or simultaneous. Simultaneous progeny choice involves multiple concurrent offspring such as a litter of mice, a brood of blackbirds, or a pod of peas. This process allows parents to screen for offspring quality among contemporaries, and allows for the replacement of defective or unsuitable offspring without undue delay. An advantage of this system is that parents can evaluate offspring quality in the crucible of sibling competition. Progeny that cannot hold their own against their siblings are unlikely to thrive in the outside world either. At the end of the process, simultaneous progeny choice allows parents to consolidate investment in a subset of their original brood with the aim of raising offspring of above average quality.

Imagine a simple behavioral rule: create $n + 1$ or $n + 2$ offspring but keep only the best n. This is analogous to a game of cribbage, in which one is dealt six cards and keeps the best four to score the best hand. The surplus offspring in broods too large become otiose when ranked alongside the others, and are discarded. Simultaneous progeny choice allows parents to upgrade mean offspring quality in a relative sense. There is no absolute threshold, as parents simply keep the best of what is at hand.

Such is not the case with sequential progeny choice. Parents who create offspring one at a time must decide with each whether to continue or withdraw parental care. The behavioral rule might be something like this: keep offspring that exceed some threshold for quality, and reject those that do not. An efficient screening system—such as might be used on a production line for lightbulbs or carburetors—may involve preliminary testing for functionality. Is this item functional or defective? Here there is an absolute benchmark of quality. Those bulbs that light pass inspection; otherwise they fail and are discarded.

Just such a system may hold in humans; in effect, mothers use a sophisticated screening system to regulate both offspring number and quality. Every potential offspring is subjected to an early performance trial, the process of implantation. Success rests on an embryo being able to turn on its genetic machinery to produce the chemicals needed to maintain pregnancy, to invade mother's uterine lining to establish its placental support system, and to evade her immune defenses. For human embryos this test poses a significant challenge, and most do not get a passing grade. Only one in two conceptions survives to term in a twenty-year-old mother, a ratio that falls to one in twenty-five in mothers over the age of forty.

Early detection of defective embryos renders the system more efficient. If reliable cues to embryo quality can be found early, mothers can make faster decisions and waste fewer resources when the decision is to terminate a pregnancy. In humans the production of the hormone human chorionic gonadotropin (HCG) may provide an early cue to embryo quality. It is a prerequisite for pregnancy maintenance and illustrates an embryo's ability to translate its genetic instructions into functional proteins, an early check of the biochemical machinery. Embryos that produce low levels of HCG are spontaneously aborted, and those that produce high levels are retained.

Progeny Choice in Humans

The mechanism of progeny choice in humans involves three steps. The first is a screening of eggs during the process of ovulation, and this naturally occurs before fertilization. The second stage occurs after fertilization, during the process of implantation. The third stage follows implantation and ends at birth. Stage one involves simultaneous progeny choice. Stages two and three, following fertilization, are sequential progeny choices.

Unlike human males, who produce sperm throughout their lives, human females are born with a complete stock of oocytes (immature eggs). Egg number is at a maximum of about 6 million during gestation, a number that falls to 600,000 at birth, 300,000 at menarche, and dwindles to only 1,000 at menopause. Of the initial 6 million oocytes, only 400 or so are actually ovulated over a woman's life. The surplus undergo a process of atresia or oocyte death. Each month, a series of oocytes are recruited from the resting pool under the influence of a surge follicle-stimulating hormone (FSH) and begin to increase in size. What the FSH triggers is a growth race among ova. But this race has a sinister dimension: the winner kills the losers before crossing the finish line. The dominant egg follicle—the largest egg in the growth race—suppresses the subordinate follicles hormonally, forcing them to commit cell suicide (apoptosis).

This growth race serves as an early test of the competence of the egg follicle and the oocyte within. Only those follicles competent to turn on their genetic machinery to complete the growth process and commence hormone production to suppress the subordinate follicles can emerge victorious. Once ovulation has occurred, the second stage of the screening process commences.

Adaptive Miscarriage

Early miscarriage is normally quick and efficient. Most embryo losses occur in the first weeks of pregnancy, either before implantation has occurred, or in the earliest stages of implantation. When miscarriage occurs this early, a mother will not be aware that she is or was pregnant. Indeed, most conceptions do not reach the stage of clinically recognized pregnancy, and for this reason we know little about these very early events, and much of what we think we know is biased. When it is possible to perform chromosomal studies of miscarriages, the rate of defects is remarkably high, particularly during the first trimester. Up to two-thirds of early pregnancy losses have recognizable chromosomal abnormalities, and an unknown fraction of the remainder have more subtle but equally lethal genetic defects. And this is only the miscarriages that we know about. It is probably a reasonable conjecture that the very earliest pregnancy losses, which almost always go undetected, have even higher rates of genetic defects, if only because losses late in pregnancy are associated with external causes, and in particular pathogens.

There is a steady attrition of genetic defects as pregnancy progresses. The rates of chromosomal abnormality among recognized

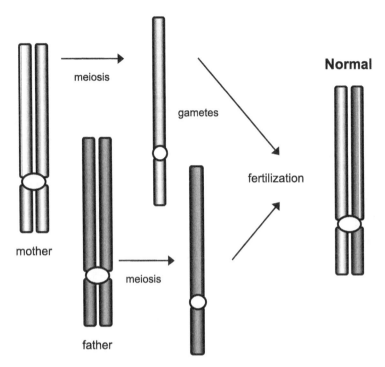

Figure 7.1. Normal sexual reproduction
In normal reproduction in sexually reproducing species, sperm- or egg-producing cells of diploid (2*n*) parents (two copies of each chromosome) undergo meiosis, a reduction division, to produce haploid (*n*) gametes (sperm or egg cells). The diploid number of chromosomes is restored at fertilization, with one maternal and one paternal set of chromosomes.

spontaneous abortuses are four to nine times higher than those observed among stillbirths, which in turn are ten times higher than those found in live births. Many human embryos are called, but few are chosen. This normally efficient system, however, falters in older mothers, as the frequency of genetic abnormalities at birth rises with maternal age. Why?

Chromosomal Defects in Humans

Humans possess 23 pairs of chromosomes, 46 in all. Twenty-two of these pairs are "autosomes," the last pair being the sex chromosomes. If all goes well, we package a single set of 23 chromosomes— one from each pair—into a gamete, an egg or sperm. This process of

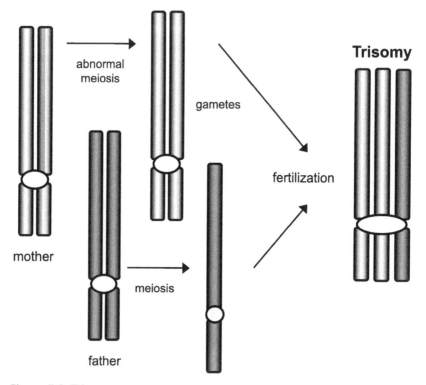

Figure 7.2 Trisomy
Occasionally during meiosis an error results in an extra chromosome being added to a gamete. After fertilization there are now three copies of this chromosome—a trisomy—instead of the normal diploid number of two. In human most trisomies result in early developmental failure and rarely if ever survive to birth; the exceptions are trisomy 13, trisomy 18, and most commonly trisomy 21, or Down syndrome.

halving the chromosomal number from $2n = 46$ to $n = 23$ is called meiosis. You and I are diploid ($2n$), but our gametes are haploid (n), so that when an egg and sperm combine, they restore the diploid number of 46 chromosomes in 23 pairs.

But the process sometimes goes awry. A tiny fraction of sperm or eggs are diploid, and after fertilization occurs, a triploid ($3n = 69$) offspring results. This abnormality is so severe that the embryo survives only very briefly. More commonly such errors in meiosis involve only a single pair of chromosomes. Consider chromosome 21, the smallest of the human autosomes. For some reason errors in assigning this chromosome correctly to gametes are more common in women than men. Some ova get an extra copy of the 21st chromosome, others none at all. When an egg with an extra copy of chromosome 21 is fertilized by a normal sperm, the resulting zygote has three copies instead of the normal two. This condition is trisomy 21, or Down syndrome. Down syndrome was originally described as "mongolism" due to a superficial similarity to people of Asian ancestry, a racist term now expurgated from the medical literature.

Down's babies possess an extra copy of chromosome 21 and thus have 47 instead of the usual 46 chromosomes. This is not good. That extra chromosome triggers a genetic imbalance with a cascade of effects: low muscle tone that affects strength, movement, and development; a nose that is smaller and flatter than normal; eyes that have a slight upward slant; a mouth that is sometimes small with a shallow roof; teeth that come in late and not in the usual order; small ears with sometimes folded tops and ear passages that may be smaller and more prone to blockage than usual, resulting in hearing loss; smaller than normal heads, hands, and fingers. Brain size and structural complexity are also reduced, and trisomy 21 children are generally mildly to moderately retarded. Forty percent are born with heart defects; 10–12% suffer intestinal problems. Down's children are also more prone to respiratory infections, vision and hearing problems, abnormal thyroid function, instability of the vertebrae, childhood leukemia, foot problems, and hernias. And adults are much more likely to develop early-onset Alzheimer's. All because of a single extra copy of a runt chromosome.

Trisomy 21 rises in frequency with maternal age. It is the most common of the three autosomal trisomies that survive to birth at measurable frequencies. Down syndrome occurs in about 1 in 800 births. Edwards syndrome, or trisomy 13, occurs in 1 in 15,000 live births, and Patau syndrome (trisomy 18) occurs in 1 in 5,000 births. What do trisomy 13, 18, and 21 have in common? They all come from very small chromosomes with at last count (National Center for Biotechnology Information Build 34 Version 1, January 2004) 463, 368, and 285 genes respectively, out of the roughly 25,000 in the entire human genome. Only the Y sex chromosome, with a meager 148 genes, has fewer. By comparison, the largest chromosome (number 1) has nearly 2,400 genes. The human genome is precariously balanced. Any insult as great as the addition of an entire chromosome will disrupt development, and the more genes on a chromosome, the more is likely to go wrong. Thus trisomy 1 is almost never seen in a clinically recognized pregnancy, as the embryo usually does not survive past the eight-cell stage of development. With chromosomes 13, 18, and 21, there are fewer genes and genetic pathways to disrupt. As a consequence these survive on occasion to birth,

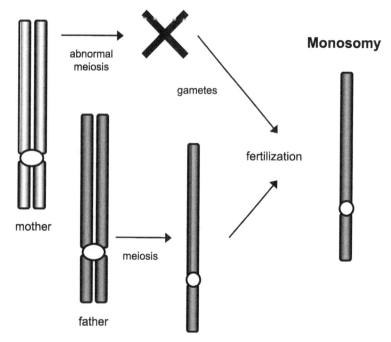

Monosomy

Figure 7.3 Monosomy
Occasionally an error in meiosis results in the absence of a chromosome in a gamete. After fertilization, one of the chromosome pairs contains only a single copy of the chromosome, resulting in a monosomy. A monosomic individual has a chromosomal complement of $2n - 1$, or one short of the diploid number. Normally a monosomy results in developmental failure of the embryo and early spontaneous abortion. An exception in humans is monosomy X, or Turner's syndrome. Here an individual is missing one of the two sex chromosomes (either an X or a Y). The single remaining sex chromosome is always an X; monosomy Y (a single Y chromosome) always results in developmental failure of the embryo.

although the odds are stacked against them. Even with the most benign and common of this trio, trisomy 21, 80% of the afflicted embryos perish before birth.

Sex Chromosomes and Birth Defects

For every extra chromosome that finds its way into one gamete, there is another gamete with a missing chromosome. When the latter combines with a normal gamete in fertilization, a monosomy results. As far as we know, monosomies are equally likely as trisomies at conception, but missing a chromosome and all the associated gene products is evidently a more serious handicap than having an extra chromosome. Monosomies almost always perish shortly after fertilization, rarely surviving long enough to make it as far as a clinically recognized pregnancy.

The only exceptions to this rule involve the sex chromosomes. Human females have two X chromosomes, males an X and a Y chromosome. There are relatively few genes on an X chromosome, and even fewer on the Y. Moreover, in women one of the two X chromosomes is rendered nonfunctional, being compressed into a Barr body early in development. This contrasts with all the paired autosomes, where both chromosomes remain functional, and because of this anomaly, it is possible to get along with only a single X chromosome.

About 1 in 10,000 females is born with one X chromosome. This is monosomy X, or Turner's syndrome. Such individuals are short, stout, and usually have nonfunctional ovaries, so that they never go through puberty. There is no overt mental retardation, though Turner's females show some difficulty in assessing spatial relationships. It is possible to survive with just a single sex chromosome as long as it is X—one cannot survive with just a single Y chromosome. The genes on the X chromosome are necessary for survival, but evidently one can do without the handful of genes on the Y chromosome.

Trisomies also occur in the sex chromosomes, and again they do not obey the same rules as the autosomes: the effects are far less severe than autosomal trisomies. Males born with an extra X chromosome (XXY), for example, develop small testes, are sterile, and are subject to hormonal imbalances. This is Klinefelter's syndrome, and in most cases there is no evidence of mental retardation. About 1 in 800 females is born with polysomy X, three copies of an X chromosome instead of the normal two. Such individuals are more prone to mental retardation, and most develop a normal reproductive system. And about 1 in 800 males possess an extra Y chromosome (XYY). Such individuals are more likely to be tall, more prone to behavioral problems, and perhaps more likely to have below-average intelligence than XY males.

Why are there more abnormalities in sex chromosomes? The reasons are threefold. First, the X chromosome is designed to work with only a single copy. (In males an X chromosome is partnered with a Y chromosome; unlike autosomes, X and Y are not homologous pairs with identical sets of genes.) Second, in females one X chromosome is inactivated when it is compressed into a Barr body; even though two X chromosomes are present, only one is used. And third, an extra Y chromosome is tolerable because Y does so little.

Turner's Syndrome and Genomic Imprinting

One sex chromosome abnormality—monosomy X, or Turner's syndrome—is particularly interesting because it appears to be linked to imprinted genes (See chapter 5). A woman with Turner's syndrome has a single X chromosome, either maternally (X_m0) or paternally (X_p0) inherited. There is no obvious effect of having a single X chromosome on outward appearance or intelligence as measured in standard tests of IQ, but there is an effect on cognitive performance, depending on whether the X chromosome was inherited from mom or dad. X_m0 individuals lack social awareness and flexibility, and perform more poorly on tests of social cognitive skills than X_p0 individuals. These results parallel those for age-matched boys (X_mY) and girls (X_mX_p): girls show more advanced social skills than boys. This genetic detective work suggests that an imprinted gene on the X chromosome affects social aptitude, perhaps via effects on early brain development. Here it is the paternal and not the maternal gene that is active. And this simply is not fair, as boys/men never get a copy of their dad's X chromosome. If they did, they would not be boys/men, who are XY; they would be girls/women, who are XX. We—and I am speaking for my sex here—get only a second-rate X chromosome from mom, one missing the good bits for social ability, and a crummy Y chromosome, which does hardly anything at all except make me male. Evidently, because of an accident of birth, I am doomed to be blissfully unaware of how much channel surfing bugs my wife.

Birth Defects and Maternal Age

There are of course more serious accidents of birth, those that arise from chromosomal abnormalities. The incidence of birth defects, and in particular trisomies 13, 18, and 21, rises with maternal age, as does curiously the incidence of dizygotic but not monozygotic twinning. Of these trisomy 21, or Down syndrome, has received the most attention. At a broad level there are two possible explanations for this pattern. Either the proportion of embryos with genetic defects rises in older mothers, or the maternal screening against defective embryos falters. These processes are not mutually exclusive, and there is indeed evidence that both are at work in humans.

An unusual feature of human reproductive biology is that every woman is born with her complete supply of oocytes, some of which eventually become the eggs that are ovulated at monthly intervals.

Human males on the other hand produce brand-new sperm throughout their reproductive lives. This disparity between men and women has lead to the stale egg hypothesis as an explanation for age-related trends in birth defects. Eggs that have been sitting on the shelf longer are more likely to go bad, and there is strong evidence for this. First and most obvious is the source of the genetic defect. A little genetic detective work shows that most of the time the extra chromosome in trisomy 21 indeed comes from mother.

Stale eggs also help to explain the decline in fertility in a woman's later reproductive years. Here the best data are derived from studies of assisted reproduction involving in vitro fertilization. The quality of ova declines with age, so much so that it becomes difficult for a woman to become pregnant after age forty. The success rate for embryo transfers using a woman's own eggs falls slightly from a woman's twenties to her early thirties, falls more rapidly until her late thirties, and falls off a cliff after about age thirty-seven. Some of this effect is due to declining uterine receptivity, but this can be reversed with the administration of supernormal levels of hormones. Most of the effect is due to declining egg quality. When eggs from younger women are used, in tandem with hormonal treatment to prepare the uterus, the success rate for embryo transfers rises dramatically.

Studies of oocytes and embryos used in in vitro fertilization also give us our best chance to evaluate the relationship between embryo quality and maternal age. And while the incidence of defective ova/embryos in older mothers rises two- to sixfold from mothers in their early twenties to those over forty, the rate of chromosomal abnormalities at birth soars twenty-five to fiftyfold. Embryo quality indeed declines with maternal age, as expected under the stale egg hypothesis, but obviously there is more to it than just that. Could it be that older mothers are less choosy?

Rejecting Low-Quality Embryos

What fraction of embryos at conception bear genetic defects? We are still not sure. This basic element of human natural history is poorly known largely because the very earliest stages of pregnancy— the first two weeks before and during implantation—are difficult to study. The proportion of chromosomally abnormal conceptions probably lies between 10 and 50%, estimates that have been derived from studies of miscarriages, early induced abortions, and embryos produced by in vitro fertilization. Whatever the number, only a small fraction of chromosomally defective embryos survive to birth.

How do mothers decide which embryos to keep and which to reject? In some cases the decision is already made. Developmentally incompetent embryos such as those with monosomies and most trisomies simply fail to implant. We might think of this as a hard screen; these embryos simply bounce off. But the hard screen does not reject all the defective embryos. What about the rest? We can also think of a "soft" screen, or one that mother can control. With a soft screen, mother gathers information about embryo quality, sets a standard for continued parental care, and aborts embryos that do not meet the imposed standard. For a soft screen to operate, mother needs to establish whether an incipient embryo is worth keeping. And here is the rub: Embryos might be tempted to fib.

During early development the control of certain tasks falls to the embryo. Embryos build the placenta, the organ that will eventually draw life-sustaining nutrients and oxygen from mother's blood supply. Embryos also produce a suite of hormones that take on a growing list of functions during early pregnancy, including keeping mother pregnant. Implantation, which occurs when the embryo burrows into the uterine wall in the days following fertilization, is a prerequisite for continued parental care in humans and represents a severe challenge to mother's immune and hormonal systems. The embryo is for all intents and purposes foreign tissue, a parasite that would normally be rejected by the maternal immune system. Thus, mother's immune system must be suppressed at the site of implantation for pregnancy to continue, and embryonic chemicals are key to this process. A variety are involved, but one in particular, human chorionic gonadotropin (HCG), is critical. Its production determines whether the embryo lives or dies.

HCG and Adaptive Miscarriage

HCG is key to pregnancy maintenance over the first eight weeks of gestation. Low early levels are associated with miscarriage, high levels with continuing pregnancy. It belongs to a family of hormones that also includes follicle stimulating hormone (FSH), luteinizing hormone (LH), and thyroid-stimulating hormone (TSH). LH and FSH regulate ovulation and control menstruation; TSH is, as you might already have guessed, involved in thyroid function. These hormones are built in two parts: alpha and beta chains that are manufactured separately and bonded together later. All members of this hormone family share a common alpha chain; thus it is the structure of the beta chain that differs and determines the biological function of the molecule. LH and HCG are particularly closely related, with LH the parent and HCG the *enfant terrible.*

In most mammals, with the exception of horses and primates, LH is the hormone of pregnancy. Mothers control whether pregnancy continues by regulating LH levels. HCG is essentially LH that is turned on in the early human embryo. The gene for the beta chain of LH and the six genes for HCG that sit side by side on chromosome 19 are virtually identical: a single mutation, a change in one of the letters in a chain of DNA, has resulted in the addition of a tail to the end of the beta chain of HCG. But this minimal genetic change has a maximal effect. It keeps the embryo from poisoning mom with its production of HCG. LH is a potent thyroid stimulator. HCG on the other hand is a weak thyroid stimulator. Indeed, if HCG had the same thyroid-stimulating activity as LH, mom would undoubtedly suffer from an overactive thyroid, a syndrome incompatible with pregnancy. But HCG, with its diminished thyroid-stimulating ability, *is* compatible with pregnancy.

HCG is important for another reason. Without it the mother's corpus luteum does not continue to produce the progesterone that maintains her uterine lining, which nourishes the implanted embryo. HCG is what evolutionary biologist David Haig refers to as an "allocrine" hormone, a hormone produced by one individual to manipulate another. In this case it is the embryo that manipulates its mother. HCG is intimately involved with the mechanism of progeny choice. For mothers it serves as an early biochemical cue of embryo quality, a cue that mothers can use for screening.

The very presence of a maternal screen creates an opportunity for subversion. Can an embryo that a mother might desire to miscarry dodge the screen? Some can and do. Certain genes related to the immune system show up far too often in offspring to be explained by chance alone: genes for child-onset diabetes, for example, appear able to dodge the maternal screen.

How do Down's embryos slip past the maternal screen? HCG appears to play a key role. Down's embryos produce a lot of it, about double the level found in a normal pregnancy. They also produce "extra strength" HCG with far greater biological activity than regular HCG. HCG is, among many other things, a miscarriage regulator. Low embryonic production of HCG is associated with a high rate of early pregnancy loss, high production with a low rate of loss. Down's embryos succeed by overproducing HCG. But even with that built-in advantage they still fare relatively poorly, as at least 80% of Down's conceptions fail to survive to birth, probably because they produce below-normal levels of other pregnancy chemicals.

Why do human mothers discard defective embryos? There is no reason in principle that the chromosomal damage could not be repaired. Instead of throwing away defective embryos, why not fix them? The reason for disposable offspring is simple. It is cheap to make new progeny. One can simply throw away damaged offspring and start anew. That said, recent work suggests that humans sometimes do repair damaged embryos.

A trisomy arises when there is an error in meiosis: when a sperm and egg combine, the resulting embryo has an extra chromosome, three copies instead of the normal two. And as the saying goes, three's a crowd. Some of these embryos subsequently undergo "trisomy rescue" at an early stage of development. Somehow, one of the trio of chromosomes is jettisoned. Imagine that there are two maternal copies of chromosome 15 and one paternal copy. Sometimes a maternal copy will be dumped, restoring the balance with one maternal and one paternal copy. But sometimes the paternal copy is lost. Now we have the correct number of chromosomes (two copies of chromosome 15), but both come from the same parent (in this case mom). This is called a uniparental disomy, and if there are imprinted genes (see chapter 5) on this chromosome, then further problems may arise. Prader-Willi and Angelman syndromes arise from uniparental disomy of chromosome 15, the former when there are two paternal copies of the chromosome, the latter when there are two maternal copies.

Relaxed Screening in Older Mothers?

Some eggs ovulated late in a woman's life have exceeded their best "use before" date, but we have reason to suspect that there is more to the higher likelihood of defects than just stale eggs. First, the problem is more than just chronological age. Aging mice show the same pattern as do humans: older moms produce more chromosomally abnormal embryos. And just like humans, mice produce eggs from two ovaries. In a clever experiment one of the two mouse ovaries was removed, showing that the process was physiological. Mouse moms now started producing aneuploid embryos (with abnormal chromosome numbers) at younger ages.

Something similar may occur in humans. A very small fraction of Turner's syndrome (monosomy X) females are not sterile but enjoy a relatively brief reproductive life span. They normally reach meno-

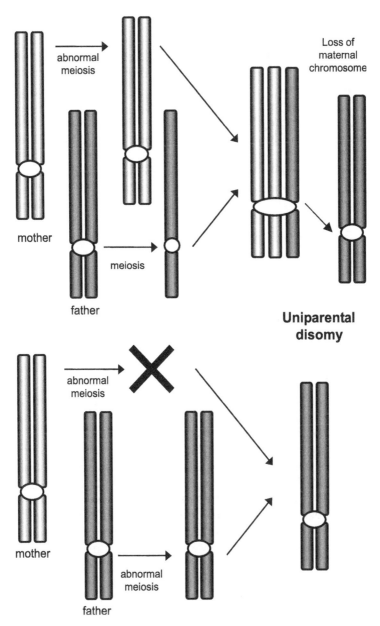

Figure 7.4. Uniparental disomy
Occasionally individuals are born with both copies of a given chromosome from a single parent. There are at least two possible pathways to a uniparental disomy. First, two abnormal gametes combine (*bottom half of figure*), one lacking a chromosome and the other diploid, to produce an embryo with the correct number of chromosomes, but both from the same parent. Second, and more likely, is trisomy rescue (*top half of figure*). By a mechanism that is not completely understood, a trisomic embryo is "repaired" by the deletion of a supernumerary chromosome. If the resulting chromosome pair includes one maternal and one paternal copy, a normal embryo results. If both copies are from the same parent, a uniparental disomy results. Uniparental disomy can disrupt the action of imprinted genes.

pause in their late twenties, show high levels of spontaneous abortion, and are particularly prone to producing trisomy 21 babies. These women grow old soon. Clearly it is a physiological, not chronological, mechanism at work. Something similar seems to happen in mothers of trisomy 21 babies: mothers with and without trisomy 21 babies were followed until menopause, and it was found that mothers of Down's babies tended to reach menopause sooner than mothers of normal babies. This result also suggests that physiological age is more important than chronological age.

The stale egg hypothesis falls short of a complete explanation for a second key reason. While there is a demonstrable decline in egg/embryo quality with advancing maternal age, the rate of increase in egg and embryo defects comes nowhere near to explaining the skyrocketing incidence of chromosomal defects and trisomy 21 in children born to older mothers, particularly in mothers over age thirty-five, a fiftyfold rise from age twenty to forty-five. Something else is going on, and the logical place to look is the maternal screening mechanism for rejecting defective embryos. This relaxed-screening hypothesis has a long and controversial history. Its critics have marshaled considerable evidence against it, noting in particular that older mothers are more, not less, prone to pregnancy loss.

The relaxed-screening hypothesis predicts that older mothers will screen defective embryos less efficiently. But all the empirical evidence from both natural and assisted reproduction (e.g., in vitro fertilization) argues otherwise. It becomes more difficult to become pregnant as mothers grow older, not easier, as relaxed selection would predict. In clinically recognized pregnancies, the proportion of gestations that end in spontaneous abortion rises from one-tenth in twenty-year-old mothers to one-third in mothers in their early forties and over one-half in mothers forty-five or older. Relaxed selection would predict the opposite. And in clinically recognized pregnancies, Down's fetuses detected early by amniocentesis fare more poorly in older mothers. If it were true that the screen operated primarily over the period of clinically recognized pregnancy, all these strands of evidence could be braided into a noose for the relaxed selection hypothesis. But progeny choice in humans operates chiefly at or before implantation—by the time that pregnancy is clinically recognized, when the embryo is several weeks of age, the game is already over. So when, in fact, does the screening break down? Part of the answer may lie in the prescreening system for ova: the process of ovulation.

Why Relaxed Selection?

Do older mothers accept lower-quality embryos? At first glance, this makes no sense. One would think that older mothers should become more, not less, selective as the frequency of defective ova rises. And in a perfect world, such would be the case. But the screening mechanism is undoubtedly biochemical. Some cue or signal is being provided by the embryo as to its quality, and mother uses this as the basis for screening. Such a decision is akin to assessing guilt in a court of law. If you want to avoid convicting innocent people falsely, you must be prepared to allow some of the guilty to go free. If you want to convict all who are guilty, you must be prepared to convict some unfortunate innocents as well. And such is the case with a biochemical screen. If one sets a rigorous threshold to screen out defective ova (the "guilty"), one also screens out many normal ova as well. But if one relaxes the screen to allow normal ova to pass (the "innocent"), more defective ova slip past. And as it becomes increasingly difficult to become pregnant, both because of an increasing frequency of defective ova and because of declining uterine function, the screen must be relaxed for an older mother to have any chance of achieving pregnancy.

The argument that relaxed screening stems from an adaptive maternal trade-off is most plausible if the screening mechanism is conservative. And it is. That is, some normal offspring are eliminated along with chromosomally abnormal offspring. Relaxing the screen enhances the likelihood of carrying a normal offspring to term but also elevates the proportion of abnormal offspring carried to term. Evidence exists to support this contention: the proportion of chromosomally normal progeny in spontaneous abortions falls with maternal age, exactly as one would expect under a conservative screening mechanism.

Experiments on strains of lab mice known to exhibit an age-related increase in chromosomal defects also point toward relaxed selection. The removal of one of the two ovaries in female mice resulted in increased ovulation from the remaining ovary. Intriguingly, this increased egg output from the single ovary resulted in chromosomal abnormalities among embryos at a younger maternal age. This result suggests that physiological, not chronological, age is more important. It is not simply that these eggs are sitting around longer—something else is going on. As in mice, so too in women. Down's syndrome is more common in mothers who have had all or part of an ovary surgically removed, or who were born with only a single ovary. Every mouse and human mother is born with a finite supply of eggs. Mothers with only one functional ovary draw down this supply much faster than mothers with two ovaries. The problem is this: the eggs at the bottom of the barrel are more likely to be defective.

Why More Spontaneous Abortions in Older Mothers?

A growing but not yet definitive body of evidence suggests that relaxed screening plays a role in the rising incidence of birth defects in older mothers. But an ugly little fact still remains. The proportion of clinically recognized pregnancies ending in spontaneous abortion rises in older mothers. If relaxed screening holds true, the proportion of conceptions that result in spontaneous abortion should fall with maternal age. A potential resolution to the paradox is suggested by work on genetic conflicts in human pregnancy (see chapter 6). If relaxing the maternal screen during an early stage of pregnancy not only reduces the incidence of spontaneous abortion but delays it as well, then the incidence of clinically recognized spontaneous abortions may rise while the overall rate of pregnancy loss falls. Here the work of David Haig is directly relevant: he argues that embryos "advertise" their vigor to the mother through the production of human chorionic gonadotropin (HCG). The proximate effect of HCG is to stimulate the corpus luteum to produce the progesterone necessary for early pregnancy maintenance. Embryos that do not meet the maternal standard—for example those with genetic defects that impair their ability to produce sufficient HCG—are aborted, before a substantial commitment of maternal resources has occurred, as evolutionary theory predicts.

HCG is probably not the only cue mothers could use, and perhaps not even the one relevant to trisomy 21, but Haig's argument suggests a potentially useful conceptual framework for the screening process and a mechanism for relaxing the maternal screen: lowering the threshold level of the biochemical cue (e.g., HCG or alpha-fetoprotein) required for continued pregnancy. For some genetically defective embryos, this would simply delay the inevitable pregnancy failure until further developmental abnormalities arose. Such a delay would shift previously "occult" abortions to clinically recognized spontaneous abortions, raising the observed rate of pregnancy loss. Other embryos with milder genetic defects (such as certain trisomies) may pass through the widened screening mesh and be carried to term.

If relaxing the screen indeed delays the loss of defective embryos, then we would expect to see not only an increase in defective embryos carried to term in older mothers but an increase in defects that result in developmental failure in spontaneous abortions (e.g., other trisomies). This in fact occurs—the incidence of most autosomal trisomies (not only trisomies 13, 18, and 21) in spontaneous abortions rises with maternal age.

The Origin of Genetic Defects

A large part of the increased incidence of birth defects in children of older mothers is due to failing gametes, as there is now no question that embryo quality declines with maternal age, particularly after age thirty-five. But mothers simultaneously relax their built-in screens in the face of this swarm of defective embryos. We can see this most directly by looking at the origin of the chromosomal abnormality in trisomy 21 births. Genetic detectives can determine whether the extra 21st chromosome came from mom or dad. If a rising incidence of defective ova is the sole cause of the increased incidence of birth defects in older mothers, then the proportion of maternal-origin trisomies should rise with the rise in birth defects. This prediction is easily testable with quantitative models. I shall skip the math and simply report the result, which is that the proportion of maternal-origin trisomies indeed rises with maternal age. This supports the notion that a rising incidence of defective ova is involved in the rising incidence of birth defects with maternal age. But the rise is too shallow, and the stale egg hypothesis is not sufficient to explain the overall pattern. It could be that older males are producing more defective sperm. This possibility has been examined but receives scant support. If there is any effect of male age on Down's births, it is very slight. What remains is relaxed screening. But when?

The Shadow of Menopause

Age brings reproductive challenges to older human mothers, chief among these simply becoming pregnant.. Fertility declines precipitously in the years approaching menopause in large part due to declining embryo quality. Younger mothers enjoy the luxury of time to be choosy; older mothers do not. A woman in her late teens or early 20s has many more opportunities to replace embryos that are rejected now; a woman in her early 40s does not. A younger mother must guard against mistakes that may jeopardize future reproduction; a mother in the shadow of menopause does not.

A conservative screening system with a significant bycatch of normal embryos alongside the defective becomes less affordable with advancing age. The problem of an increased incidence of defective ova poses special challenges for older mothers. If the mesh of the built-in screen is unaltered, then the rise in birth defects will exactly match the two- to sixfold rise in defective embryos from a woman's 20s to her early 40s. Women could compensate for a rise in embryo defects by tightening the screen, choosing in effect a smaller mesh. Doing so would have the advantage of reducing costly birth defects but would also have the undesirable effect of increasing the bycatch of normal embryos. Fertility would decline even further, and the likelihood of successful pregnancy would become more remote. This is probably why older mothers do not tighten the screen; rather they do exactly the opposite.

Older mothers increase the mesh size, allowing more defective embryos through but, more importantly, allowing more normal embryos through as well. And we must remember that even in older mothers birth defects are rare, normal births common. From age twenty to age forty, the risk of a trisomy 21 birth rises from 1 in 1,600 to 1 in 100. If we include all chromosomal defects, the risk is 1 in 60 births at age forty. Even though the risk of birth defects has in relative terms soared, they are still rare even at age forty. At any reproductive age, mothers most of the time give birth to chromosomally normal babies. As mothers age, their built-in screen is relaxed gently, a necessary compromise in the face of declining fertility. But this is not all that is going on in human mothers with the advance of age. The rate of dizygotic twinning rises alongside the incidence of birth defects until a woman reaches her early forties. There is reason to suspect that this too is part of a progeny choice mechanism.

Screening, Maternal Age, and the Role of Genomic Imprinting

For most genes in most organisms, it matters not whether the gene was maternally or paternally inherited. But for a small subset of imprinted genes (about four dozen in humans) it matters a great deal (see chapter 5). Many of these genes are involved in regulating embryonic or fetal growth, over which there may be conflict between genes of maternal or paternal origin. Under David Haig's genetic conflict hypothesis, paternal genes should be selected to extract more resources from mother than maternal genes if the mating system is other than perfect monogamy. Simply, a maternal gene in an embryo is equally related to all of its mother's future children, but such is not the case for a paternal gene if mother mates with different fathers. This results in conflict between some of an embryo's maternal and paternal genes, and such genes are expected to influence the level of maternal investment in an offspring conditional on their parental origin. Here we argue that the level of conflict, and the optimal level of maternal investment, will also be conditional on maternal age.

The genetic conflict hypothesis rests on a trade-off between current and future offspring; investing more in current offspring, as paternal genes might prefer, reduces the available investment, for future offspring, an option that maternal genes might prefer. But in humans, expected future reproduction declines with maternal age, being maximal at the age of menarche and falling to zero at menopause. Thus the value of future maternal siblings to paternal and maternal genes in an embryo changes with the mother's age. The optimal level of investment in current offspring is therefore conditional not only on the parental origin of genes influencing resource allocation but on the mother's age as well, and this should affect the extent of genetic conflict. With a diminishing potential for future reproduction, the optimal investment in current offspring should rise with maternal age, reaching a maximum in a mother's final reproductive bout before menopause. And here genetic conflict should vanish, because there is no future reproduction to contest. At this point the investment in current offspring should be that which maximizes the fitness of current offspring, a view shared by both maternal and paternal genes. If this logic is correct, then maternal resistance to paternal genes should wane in older mothers. Thus one should expect relaxed imprinting in older mothers, with increased benefits for the embryo at the mother's expense.

It is tempting to suggest that is the case in humans. Three phenomena associated with pregnancy immediately suggest themselves: trisomy 21, gestational diabetes, and preeclampsia. Trisomy 21 rises in frequency with maternal age. It is associated with a high production of an active form of human chorionic gonadotropin (HCG), which regulates the rate of miscarriage, and there is strong circumstantial evidence that HCG is affected by imprinted genes (see chapter 5). Gestational diabetes is associated with the imprinted IGF2 gene and its receptor, and represents an embryonic manipulation to aid fetal growth by elevating levels of circulating blood sugars. It too rises in frequency in older mothers, as does preeclampsia or pregnancy-induced hypertension, which permits embryos access to more maternal resources. Are these examples of reduced genetic conflict in the embryos of older mothers? Perhaps, but it is too soon to tell.

Maternal Age and Twinning

One of the incentives for creating surplus incipient progeny under a strategy of parental optimism (chapter 2) is to expand the pool of candidates for progeny choice. The likelihood of uncovering a high-quality offspring will grow with the initial brood size. With a decline in the quality of ova and embryos with age, might it make sense to create a larger brood from which to choose? In humans with a modal brood size of one, the opportunities for simultaneous progeny choice would seem limited. But close inspection reveals that human mothers do integrate this component of parental optimism into their reproductive strategy. It occurs via polyovulation, the production of more than one egg at a time, and gives rise to dizygotic twins, which along with birth defects also increase in frequency with maternal age. Women over age thirty-seven begin to show a different pattern of hormone production during their monthly cycle. In younger women the levels of follicle-stimulating hormone (FSH) rise in the first two weeks of the cycle and fall thereafter. In older women FSH levels do not fall off as quickly. This opens the door to polyovulation, and explains the rising incidence of dizygotic twinning in a woman's late thirties. But does it also open the door to defective ova? I shall turn to this question next.

Chapter 8

Why Twins?

The Evolution of Brood and Family Size

By the standards of Victorian England, Charles and Emma Darwin led lives of privilege. They received £18,000 as wedding gifts from their parents and a further £45,000 inheritance with the death of Charles's father Robert in 1848. By 1882 shrewd investment had more than quadrupled the estate to £282,000. Their independent wealth allowed the Darwins to live a quiet life and raise a large family in the village of Down. Freed from the pedestrian concerns of having to earn a living, Charles Darwin was able to devote himself to a life in science. Even before the publication of the *Origin of Species*, he had risen to the elite ranks of British science, receiving the Copley medal from the Royal Society in 1856 for his work on geology and invertebrate zoology.

Over the same years Emma was busy having children. Even the comparative wealth of the Darwins, though, could not exempt their children from the same grim fate of so many other contemporary families. Emma gave birth to ten children between 1839 and 1856. Of these only seven survived to adulthood. Mary Eleanor Darwin died just three weeks after her birth in 1842. Charles's favorite, Annie, died tragically after a protracted and mysterious illness at age ten, with symptoms that paralleled those of her father. Her death extinguished Charles Darwin's Christian faith. And in June of 1858 eighteen-month-old Charles Waring Darwin, sickly from birth, contracted scarlet fever. On the eighteenth of the same month a package arrived from halfway across the globe, and in it was a manuscript from the young naturalist Alfred Russel Wallace. Wallace had independently derived Darwin's own, but as yet unpublished, theory of natural selection, the central focus of Darwin's scientific life over the last two decades. On June 28, after a brief recovery, baby Charles died. A grief struck the father; the life of his son had just been stolen from him, and now too, perhaps, credit for the most important scientific discovery of the nineteenth century. Darwin turned the matter over to his friends and colleagues, geologist Charles Lyell and botanist Joseph Hooker. They arranged for a simultaneous presentation of manuscripts by both Wallace and Darwin at the meeting of the 1858 Linnean Society of London. The following year, and after an extraordinary gestation, Darwin published the *Origin of Species*.

In mid-nineteenth-century Europe three of ten babies perished in infancy. A century and a half later that rate has plunged to fewer than one in a hundred thanks to improved nutrition, sanitation, and advances in medical science. But not all share in first-world prosperity. In rural populations across the world infant death rates still soar. At the beginning of the new millennium nearly two in five babies die in their first year in the poorest countries of the world, more still before the children reach five years of age. In the year 2000, 26% of children in Afghanistan died before their fifth birthday, in Niger 27%, in Angola 30%, and in Sierra Leone 32%. The comparable figures for the United States, Canada, and the United Kingdom were 0.8%, 0.6%, and 0.6% respectively, with Japan, Iceland, Norway, Singapore, Sweden, and Switzerland leading the world at a minuscule 0.4%. Wealth confers privilege.

We in the industrialized West have been largely emancipated from the natural checks on health and fertility, and many of the legacies of evolutionary history have been rendered impotent. Mothers can give birth to babies in close succession with barely detectable effects on the health of mothers and babies. But for most of humanity, and for most of our evolutionary history, it is not (and was not) always so. Historical records from German noble families for example, dating back to the sixteenth century, show that 9% of newborns died in their first week. But that proportion had fallen to just 0.5% by the early twentieth century. One out of 5 sixteenth-century babies failed to survive the first year, compared with one of every 125 babies between 1900 and 1930. Swedish data show the same pattern, a drop in infant mortality rates from 1 in 5 in the late eighteenth century to just 1 in 143 two centuries later.

In rural Finland two centuries ago twin births resulted in about the same number of surviving children as singletons, whereas today with advanced medical care the infant mortality rate for twins is only slightly higher than that for singletons. Twin births now result in roughly twice as many surviving babies as singletons. This pattern should come as no surprise. Improvements in nutrition, sanitation, and the medical sciences all contribute to the plummeting rates of child mortality for all of us in the industrialized West. We now expect our babies to survive. But measured in evolutionary time, this is a recent innovation. Humans are designed to expect failure.

Fault-Tolerant Design in Humans

Avian reproduction or mammalian gestation is much like a space-flight. Once it has begun there is no turning back. Failures must be anticipated, with backup systems in place, or the mission will need to be aborted. Compare this to a large seabird laying a single egg. After ten weeks of incubation it finds that the egg is a dud, and it is now too late in the breeding season to begin anew. The obvious solution? Lay a second, backup egg that can be substituted for the first should the worst happen. This is fault-tolerant design, and it allows parents to compensate for failure.

Computer scientists and aerospace engineers routinely plan for failure by implementing fault-tolerant design. One anticipates accidents or defects in critical systems and devises strategies to compensate or adapt. This may involve incorporating redundancy in the original design or planning for a degraded mode of operation when acceptable. Thus when a bulky high-gain antenna on a Mars lander is knocked out of service by fiendish martians, a backup medium-gain antenna is used in its place. It is still enough to get by, even though the rate of data transmission is reduced. Fault-tolerant design is not without its own costs: there are costs associated with the construction and maintenance of extra elements that may be called into service as backups, and there are problems associated with additional system complexity.

Fault-tolerant design is also a routine component of reproductive behavior in birds and mammals. Brood reduction is an obvious example. When the input of resources falls short of that required to sustain the full brood, parents have a built-in system to shed low-priority, marginal offspring in favor of high-priority, core offspring. Doing so allows parents to consolidate investment in a smaller number of higher-quality progeny. Similarly, on a spacecraft, when a solar panel array fails, diminishing the electrical supply, a built-in software program shuts down low-priority electrical circuits and instruments while maintaining high-priority mission capabilities. For birds, the costs of fault-tolerant design are the extra costs of building and maintaining eggs that may not be needed. And if the eggs eventually prove surplus to their parent's needs, what will be done with them?

Humans also build redundancy into their reproductive strategy, both before and after birth. Postnatal strategies involve what human population demographers call "insurance" and "replacement." They define these terms in a different manner than I use them here, but the broad themes are similar. Insurance (aka hoarding) is the creation of a larger than optimal family in the absence of premature offspring mortality: the surplus offspring are created as a hedge against possible offspring loss. By now it should be evident that this is a strategy widely shared with many plants and animals. Replacement is a strategy to compensate for actual infant mortality, and involves an abbreviation of the interbirth interval following the death of a child. Parents can compensate partially for the loss of an earlier child by accelerating the birth of the next child. Both insurance and replacement strategies are widely observed in humans in rural societies across the world.

Twinning as an Insurance Strategy

Redundancy is even more important before birth in humans. We call it twinning, and it is far more common than we might suspect. The logic of redundancy is that a substitute can be called into place if a critical component fails, allowing the system to function with minimal interruption. Multiple births are rare in humans, but multiple conceptions are not. As many as 1 in every 8 pregnancies begin as twin conceptions, though few survive intact to birth: twins constitute only 1 in every 80 to 100 live births, and many twin conceptions fail entirely. If twins rarely survive, why then are they so common at conception? The answer is insurance.

Most conceptions, twins or not, do not survive to birth. Most do not even survive until pregnancy is clinically recognized, and as mothers grow older, the survival rate for these embryos falls even further. Twin conceptions enhance the likelihood of successful fertilization and implantation, but at a potential cost: twins, or more specifically twin births. And here I want to focus on the type of twins—dizygotic—that is under maternal control. Dizygotic twins arise from two separate eggs ovulated by a mother during a single monthly cycle. The resulting offspring are no more closely related genetically than ordinary brothers or sisters. Monozygotic twins arise from the fission of an already fertilized egg (i.e., one zygote) and appear to arise chiefly as an accident of early development. If this is indeed true (and its truth is a subject of some discussion), then mothers really do not "choose" to have monozygotic twins in the same sense that they "choose" to have dizygotic twins by polyovulation.

Are dizygotic twins in humans adaptive? The answer to this question is not at all clear. Over the course of much of human evolutionary history, the answer is probably no—mothers for the most part would have been better off avoiding twin births. But at the same time, mothers benefit from polyovulation, and perhaps twin conceptions, because they increase the likelihood of successful pregnancy.

Insurance Offspring in Birds

The notion that twinning is part of an insurance strategy was first suggested by the behavioral ecologist Dave Anderson, who normally works on large seabirds—boobies—in the Galapagos. Some boobies, along with certain pelicans, cranes, eagles, and penguins practice obligate brood reduction (see chapters 2 and 10): they lay two eggs, but if both hatch, one chick—usually the younger—is a target for infanticide either by parental neglect or sibling aggression. Parents lay two eggs to enjoy the advantage of redundancy: insurance. All these birds have long incubation periods; parents that waited until hatching time only to find that their lone egg was inviable would likely forfeit an entire breeding season. Better to add a relatively cheap insurance egg at the start than to risk losing a whole year.

The first critical test of the insurance hypothesis was conducted by behavioral ecologists Roger Evans and his student Kevin Cash, working on white pelicans in southern Manitoba. These pelicans are typical obligate brood reducers that normally lay two eggs but rarely raise two nestlings to independence. Instead, when both eggs hatch—several days apart—the older nestling bludgeons its younger nest mate into an early demise. Why two eggs if only a single chick ever survives? The experiment was elegant and simple. Evans and Cash removed one of the two eggs shortly after laying and compared the fate of these clutches with the fate of those in which the two-egg clutch remained intact. The clutches of two yielded more chicks than clutches artificially reduced to one, even though no more than a single chick survived in any brood. The reason was that about one in five eggs failed to hatch. Thus, parents with a clutch of one failed to rear any offspring 20% of the time. But with a second, backup egg present the rate of complete failure fell from one in five to one in twenty-five, the latter number being the probability that both eggs would fail to hatch. The marginal chick was expendable and eliminated from the brood by siblicide if both eggs hatched.

Other workers have since replicated these results not only in other obligate brood-reducing birds but in facultative brood reducers too. Red-winged and yellow-headed blackbirds, residents of prairie wetlands, typically lay three to five eggs and hatch them asynchronously. The last-hatched, marginal nestlings suffer much higher mortality than the first-hatched, core nestlings. But when core nestlings are removed experimentally, or core eggs fail to hatch in un-manipulated nests, the survival rate of the marginal nestlings soars. Part of the value of these marginal progeny is as insurance against the premature failure of core offspring. A key component of this insurance strategy is that the marginal progeny are low-cost offspring.

In Vitro Fertilization and Twinning

Our knowledge of the frequency of polyovulation and the survival of each conceptus in natural human pregnancy is scant, and much has to be inferred by mathematical modeling. But the new science of assisted reproduction, which began in 1978 with the first test-tube baby, offers a window on the events of early reproduction. In vitro fertilization involves the artificial stimulation of a woman's ovary hormonally, a harvest of unfertilized eggs, artificial fertilization (usually in a petri dish instead of a test tube), and replacement of the fertilized eggs into the woman's uterus. As the exact number of fertilized eggs that are transferred is known, a number that varies from one to seven or more, so too is the survival rate of embryos. And what we know is this. In mothers under thirty, the success rate—transfers that result in a live birth—for a single-embryo transfer is about 10%. This rate falls gradually throughout a mother's thirties and plummets after her fortieth birthday. The success rate for single-embryo transfers in mothers over forty who use their own eggs is a discouraging 2%. But the take-home baby rate rises with the number of embryos transferred. For mothers under thirty the success rate rises from about 22% for two embryos to nearly 40% for three embryos. These numbers also fall with maternal age, as with single-embryo transfers, and the fact that take-home baby rates are higher when more embryos are transferred has not escaped the notice of IVF clinicians. Larger numbers of embryos are routinely transferred for mothers over the age of forty, raising the take-home baby rate from 2% for a single embryo to close to 20% for five or more embryos.

Of course when multiple embryos are transferred, multiple pregnancy often occurs, the record being twelve embryos. But this risk of multiple pregnancy falls with maternal age. When three embryos are transferred to women in their forties, only 11% of live births will be twins or triplets, but for mothers under thirty that frequency soars to nearly 50%. Thus the risk of multiple gestation is comparatively low in older mothers. But if this is so, why then are dizygotic twins *more* common in older mothers?

The answer is simple. Older mothers polyovulate more often than younger mothers. Way more often. Again there are few direct observations of the frequency of polyovulation in natural pregnancies, but we can use the IVF data as a guide. If we assume that embryo mortality in natural pregnancy is roughly the same as that observed in IVF pregnancies, then perhaps one in thirty-five gestations begin as twins in mothers under the age of thirty, rising to one in eight in mothers between thirty-five and thirty-nine. These rates might be a bit high, as the mortality of embryos in IVF pregnancy might be higher than in natural pregnancies, but the key point is this: polyovulation enhances the chance of achieving a successful pregnancy in older mothers. Most of the time only a single embryo survives even when three eggs are fertilized. Twin births in natural pregnancies in older mothers are in fact the small tip of a very large iceberg of twin conceptions, and for good reason. Twin pregnancies are dangerous.

Twinning is not about twins. It is about avoiding pregnancy failure: the failure for an egg to be fertilized, the failure of a fertilized egg to develop normally, and the failure of a fertilized egg to implant in the uterus. And twinning is also about *avoiding* twins. Mothers risk dizygotic twinning only when twin births are unlikely, or when they are better prepared for the steep costs of twinning than most. Thus, the incidence of polyovulation, and hence the risk of twins, rises as mothers age and embryo survival falls. Just as in obligate brood-reducing birds, whose two-egg clutches are about producing one viable offspring, twin conceptions are about producing a single healthy baby.

Mothers avoid twins because twins are both expensive and risky. Feeding and caring for two fetuses/infants at a time costs more, and the risks associated with twin gestations are high for both mother and babies. A twin pregnancy begins with elevated levels of first-trimester nausea and vomiting, and by the end of the third trimester mom has added an extra forty to eighty pounds of fat in anticipation of the heavy costs of lactation. In between, a pregnant mother suffers an elevated risk of preeclampsia (pregnancy-induced high blood pressure), gestational diabetes, and hemorrhage due to *placenta abruptio* (when the placenta tears away from the uterine wall) or *placenta previa* (the placenta covers the cervix). There are elevated risks at delivery, so much so that the majority of twins in the United States are delivered by caesarian section. And even with modern medical care, the risk of maternal mortality is many times higher than that for singleton pregnancies.

The statistics are even more grim for the twins themselves. Given that as many as one in eight gestations begin as twins, but only one in eighty end as twin births, the math is simple. Most twin gestations undergo brood reduction, losing one and often both embryos. Most of this brood reduction occurs early and out of sight, before the embryos technically become fetuses at eight weeks of age: these are "vanishing" twins. Some but not all of these pregnancy losses can be detected with ultrasound examination.

The mortality statistics differ according to whether the twins come from one or two eggs, and how the placenta and amniotic sac is constructed. Monozygotic twins sometimes share a fused placenta (monochorionic). As an embryo grows, it is enveloped by a fluid-filled sac, the amnion. Some monozygotic twins share a single sac (monoamniotic); others live in separate sacs (diamniotic). Twins that have the misfortune to share a single sac and a common placenta have a staggering mortality rate, and if they do survive, they experience very high rates of congenital defects such as cerebral palsy. Monochorionic/diamniotic twins fare somewhat better, and dichorionic/diamniotic twins best of all. Dizygotic twins are always of this last type, and thus on average fare much better than monozygotic twins.

The sublethal effects of twinning are also significant. Twins grow slower in the third trimester of pregnancy, and are usually born earlier and smaller. Twins of either type are more prone to cord accidents at birth, and are far more likely to be born with physical and mental disabilities than singletons. The risk of cerebral palsy is tenfold higher in twins than in singletons due to oxygen deficits during gestation that injure brain development.

Dizygotic twinning, however, is heritable, and if twinning was maladaptive we might expect natural selection to weed out such a trait. But twins per se are only a small part of a larger strategy designed to enhance the fertility particularly of older women. The chief benefit derived from polyovulation is an increased likelihood of a successful singleton pregnancy. Twin gestations and births are potential costs, though the latter are relatively rare. What is not clear is whether mothers with twin conceptions adopt a strategy of deliberate brood reduction—in utero—to avoid the costs of twins, and if so, how. That may well be the ideal strategy. Begin with an optimistic brood size and trim the surplus as conditions warrant—the standard strategy of parental optimism that is so widespread in nature. It may be that mothers who can afford twins—taller, fatter individuals in good physiological condition—are the ones who carry twin gestations to term. The alternative is also possible: mothers polyovulate and then let nature take its course. Most of the time that would result in either no babies at all or a singleton pregnancy. In a small proportion of cases, twins arise.

Age, Trisomy 21, and Twinning

Anthropologists Helen Ball and Catherine Hill have identified a particularly intriguing correlation in a survey of human populations. Twinning appears to be associated with an elevated risk of trisomy 21 (Down syndrome). They suggest that mothers polyovulate as a hedge against defective embryos, in this particular case against embryos with the wrong number of chromosomes. This hypothesis, though not yet proven, has a lot going for it. It explains not only the population data but also the rising incidence of both dizygotic twins and trisomy 21 births with maternal age. It also explains two otherwise paradoxical patterns. First, trisomy 21 babies are overrepresented among twin births. Though the data are scarce, trisomy 21 occurs more often in dizygotic twins pairs than one would expect by chance alone. Second, mothers prone to twinning are also prone to trisomy 21, though not necessarily at the same time. Hill and Ball's hypothesis, an extension of David Anderson's earlier work, tidily explains these patterns, and the fact that the incidence of dizygotic twinning rises sharply after age thirty-five to a peak around age forty and then falls off thereafter due to declining ovarian function and/or embryo quality.

But could there be another explanation? Perhaps it is that mothers prone to twinning exhaust their pool of high-quality embryos sooner and hence are more prone to trisomy 21 embryos later in life. There are two suggestive pieces of evidence. First, mothers prone to twinning—supermoms—tend to be more fertile. There is a birth order effect in this pattern: the more previous children a woman has had, the more common twins become, independent of age. It is probably not birth order per se that is important; rather it is the hyperfertile women—supermoms—who get to these high birth orders.

Second, there is weak evidence that mothers prone to twinning go through menopause sooner, suggestive of ovarian exhaustion. Chang-Jiang Zheng and Breck Byers have suggested an intriguing notion concerning the origin of trisomy 21: oocyte selection. They posit that when a woman initiates multiple egg-producing cells—oocytes—each month to ovulate a single mature egg, normal cells are more likely to win this race. As women grow older, the stock of normal oocytes dwindles, and the likelihood of defective eggs (ova) slipping though this early screen rises. This could in fact be the basis of the maternal screen for defective embryos (see chapter 7). If Zheng and Byers are correct, then it would follow that supermoms would reach menopause sooner, and dip into the pool of defective ova earlier, providing a potential explanation for Hill and Ball's correlation between trisomy 21 and dizygotic twinning. Unfortunately, we still lack the necessary information to test these ideas directly. Surprisingly little is known about many basic elements of the natural history of human reproduction. How many eggs are in fact ovulated during a monthly cycle? How much does this vary across women and with maternal age? There is still much to learn, and only now are these data on fundamental aspects of human natural history beginning to be collected with the assistance of modern technology.

Is it that defective oocytes are less likely to ovulate than normal oocytes, as Zheng and Byers suggest? This provides a tidy explanation for the age-related increase in birth defects with advancing maternal age. But the proximate mechanism is not known. The process of ovulation in fact lasts several months, and it is only those egg follicles that have reached a threshold size that participate in the monthly growth race. What is it that triggers a resting egg follicle to begin growth at the very earliest stage? And is this linked to egg quality? The Zheng and Byers hypothesis is attractive, and may well be verified as we learn more about the early events of ovulation.

But events at the other end of ovulation could play a role. Older mothers produce more follicle-stimulating hormone (FSH), and the result is that more than one egg may survive this prescreening phase of human progeny choice. By allowing more ova through the pre-screen, do mothers also let more low-quality eggs through as well? The rise in incidence of dizygotic twinning may signal when a mother has relaxed her built-in screen, and thus explain the rising incidence of birth defects in older mothers. Or, if Zheng and Byers are correct, it is not that mothers are relaxing the screen but rather that, toward the end of a woman's reproductive life, the proportion of defective ova in the remaining pool of egg follicles rises sharply. Herein lies an intriguing question. Does polyovulation cause more defective eggs to be let through, or does the rising incidence of defective eggs cause polyovulation? The direction of causality is unclear, and both mechanisms are compatible with an insurance strategy.

More Than Just Polyovulation

That dizygotic twinning is heritable has long been known. It tends to run in families, and daughters of mothers of twins are much more prone to multiple births than daughters of mothers of singletons only, suggesting that polyovulation is heritable, though there is surprisingly little direct evidence. But there is more to it than just polyovulation. There is a paternal link to twinning. Recent work on tracing the incidence of twinning in Scottish families has revealed a startling pattern: twinning runs in families but via fathers, not mothers (i.e., a paternal transmission of twinning). That there is a maternal transmission of twinning is well established, and the mechanism is easily understood: an increased tendency toward polyovulation. But how might a paternal transmission work? The answer is stunning and challenges much conventional thought about twinning. Human geneticist Mikhail Golubovsky suggests that some human males produce unusual sperm that yield double fertilizations, which in turn produce a triploid zygote. Triploidy means that there are three copies of the genes instead of the normal two, with double copies of the paternal genes. This represents a severe genetic abnormality, and one that is not normally compatible with continued development.

Triploid zygotes are not uncommon, comprising roughly 1% of human conceptions and more than 10% of spontaneous abortions. Many of these are inviable and spontaneously aborted, but some, evidently, are not. Triploidy appears to be a transient state and is associated with developmental instability. There are at least three different developmental pathways for a zygote that finds itself with an extra set of chromosomes.

The first is to remain triploid ($3n$), giving rise to partial moles (see chapter 5) that have a double set of paternal chromosomes. These moles are not viable and potentially dangerous to the mother. The second pathway involves "diploidization" of the triploid zygote, yielding some cells with a diploid complement ($2n$) and other cells with either a haploid ($1n$) or triploid ($3n$) chromosomal complements. The haploid cell(s) may restore the diploid condition by duplicating their single chromosome set. This is good and bad, as diploid is better than haploid, but both sets of chromosomes are from the same parent, which is bad. This second pathway can result in a variety of outcomes: individuals that are a mosaic of different cell types; individuals that are diploid or mosaics of diploid cells with different parental origins; complete moles; or fetus/mole tandems. Individuals that are cellular mosaics are particularly interesting. Cells taken from different parts of the body could potentially show different genetic fingerprints. The mind boggles at the possibilities for television crime dramas.

The third developmental outcome arises from developmental instability associated with triploidy. Embryos developing from triploid zygotes may be pruned back to an approximately diploid ($2n$) chromosome number, but with an extra chromosome here and there

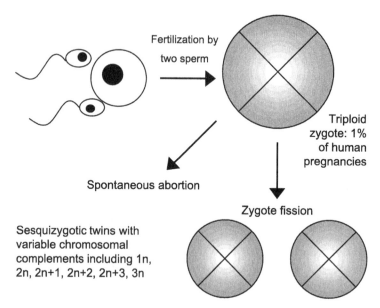

Figure 8.1. Sesquizygotic twins?
Twinning sometimes shows a pattern of paternal transmission that is inconsistent with convential monozygotic or dizygotic twinning. This third pathway is referred to as sesquizygotic twinning. About 1% of human conceptions are triploid, arising most often from the fertilization of an egg by two sperm. The resultant zygote appears to be developmentally unstable and prone to fission, with a wide array of potential chromosomals complements (haploid, diploid, triploid, partial and complete moles, single or multiple trisomies) in the resultant embryos. This mechanism potentially explains the paternal link to twinning and the link between twins and chromosomal defects.

(one or more trisomies), or may give rise to uniparental disomy (both chromosomes from a single parent—see chapter 7 for discussion of disomies and trisomies). All three pathways may also lead to embryo fission, resulting in twins or fetus/mole tandems. This represents a third pathway to twinning referred to as "sesquizygotic." If Golubovsky is correct, it may explain the link between twinning and a variety of developmental abnormalities, for example the tendency of twins and trisomy 21 to be linked in some families.

The paternal link to twinning highlights how error prone are the basic components of human reproduction, such as the production of gametes, fertilization, and early embryonic development. Or perhaps some of these may not be errors at all but something more sinister: genetic intrigues on behalf of paternal genes. One possibility is that if males can induce mothers to bear twins by tinkering with the mechanics of fertilization and early development, they can potentially extort greater parental investment from females. This seems extremely unlikely given the adverse outcomes that are likely, and given that even when phenotypically normal, twins rarely outperform singletons. More plausible is a battle of the sexes that has as a by-product an increased likelihood of dispermy.

Dispermy is the fertilization of an egg by more than one sperm, with triploid embryos being the frequent result. In humans there are two built-in mechanisms to restrict the entry of more than one sperm into an egg. The first is the rapid block to polyspermy that occurs within one to three seconds of the sperm contacting the outer jelly coat of the egg. It is a rapid depolarization of the plasma membrane surrounding the egg and prevents more than one sperm fusing with the plasma membrane. The second is the slow block to polyspermy. It occurs via the release of calcium, which activates cortical granules that lie just beneath the cell membrane of an egg. The contents of these granules fuse with the membrane and induce the development of a spermproof fertilization membrane. The reaction starts at the site of sperm penetration and spreads as a wave over the entire cell, a process that takes twenty seconds or so to complete.

Sperm and eggs face differing selection pressures across evolutionary time. For an egg, the goal is to avoid the disaster of dispermy. For a sperm, there is no reward for being the runner-up. Being first is what matters. Thus, while eggs build defenses against multiple fertilization, sperm develop methods of penetrating eggs rapidly. And this need for speed may inadvertently disable the fast block to polyspermy. Here maternal and paternal genes face conflicting selection pressures, and a mechanism to dodge maternal defenses would face strong opposing selection. Molecular biologists provide support for this view, showing that genes influencing fertilization undergo particularly fast evolution. And in evolution, as in car chases, excess speed can have unwanted by-products: a paternal link to an aberrant form of twinning may be one such accident.

Twinning and Individual Optimization

Population ecologists have long been puzzled by the fact that birds lay clutches of varying sizes, and many seem to lay fewer eggs than they can rear in a given year. Life histories are shaped by trade-offs. One key trade-off is the principle of allocation. Resources are finite and compromises necessary. Long ago, the British ornithologist

David Lack pointed out that parent birds must choose between rearing more low-quality offspring or fewer high-quality offspring. As natural selection favors leaving more descendants, rearing the most nestlings may not pay if they leave the nest malnourished and perish soon after. Better to provision a few offspring well than a larger number of runts. Lack suggested that the optimal clutch size is that likely to leave the most surviving offspring. But population ecologists have long been puzzled by the fact that parent birds seem to lay fewer eggs than the Lack clutch size. There are many possible explanations for this discrepancy, and not all are mutually exclusive. One that is likely generally true, though difficult to show, is that parents hold something back to protect future reproductive opportunities—the cost of reproduction hypothesis. A lower investment in current reproduction saves more for future reproduction, an important issue in long-lived species whose breeding is a marathon, not a sprint.

But the choice of family size is also influenced by differences across individuals. Some parents, quite simply, are better than others. Some are more experienced or in better health or nutritional condition than others. Some, such as large-billed Galapagos finches during times of drought, are serendipitously better suited morphologically (and genetically) to ephemeral local environments. And some parents differ in opportunity. Some live on better territories, or have acquired the social status of their parents.

Darwin based his concept of population thinking, a key component of his theory of natural selection, on these differences among individuals. It is a concept we still struggle with a century and a half later. A very different philosophical view—typological thinking—has dominated Western thought for the last two thousand years, dating back to Plato and Aristotle. Under typological thinking we classify objects into types or groups. Here the essence is defined as important, and the differences across objects within the same group are trivial. We can talk of classes of objects, such as cars or rivers or fences, and ignore the differences between, say, a Bugatti T57 and a Dodge Dart, or between the Berlin Wall and my backyard fence. For the physical sciences of chemistry and physics, typological thinking works quite well. A hydrogen atom is pretty much the same on Earth as it is on Alpha Centauri (we think). But typological thinking just will not do for the biological sciences. The differences across individuals within species and populations matter. If not for the diversity within populations, we would not have the diversity of life itself, for it is the raw stuff of evolution. And describing diversity has always been problematic for biologists. Our manner of describing species was founded on typological thinking: when biologists describe a new species, they designate a "type" specimen, meant to be representative of the species' characteristics. This concept dates back to the great Swedish botanist Carolus Linnaeus, who established our system of naming and classifying species. Linnaeus was a creationist, as were all eighteenth-century biologists. To him, the modern concepts of variation and evolutionary change would have been unthinkable. Typological thinking is deeply embedded in the sciences, biology being no exception. Even ecologists often describe species and populations by their average properties—an average clutch size, a typical mating system, a normal behavior—instead of focusing on the variation within species and populations. We ignore these differences at our own peril.

Take almost any biological population at a given time and place, and you will find within it myriad differences across individuals. Some are obvious—differences between the old and young, male and female—but others are less so. Two individuals that appear outwardly identical may differ in important ways. In a population of red-winged blackbirds in the prairie wetlands of Manitoba you will find some females who lay only two eggs, others who lay six—most lay three to five eggs. Those laying smaller clutches tend to be younger than those laying larger clutches, and younger birds are often less experienced, less efficient at gathering food, and less competitive with other members of their own species. They often skip breeding entirely or, when they do breed, raise smaller families. Variation across individuals also arises because some carry heavy loads of parasites or pathogens, leaving them less able to raise families. Still others may have been less lucky in securing food over the winter months and arrive on the breeding grounds in poor physiological condition. And some may secure access to better breeding territories than others due to superior fighting skills.

Just as in humans, some avian families are blessed with more. They can afford better cars, bigger houses, larger nests, and more eggs. Experience and access to resources also make a difference, a phenomenon that has been closely studied in breeding birds and formalized as the individual optimization hypothesis. An elegant series of experiments conducted on great tits nesting in Oxford's Wytham Woods illustrated this effect quite neatly. When the population as a whole was examined, the average clutch size was smaller than the clutch size that yielded the maximum number of surviving offspring. Parents seemed to be conservative in setting their family size. But when individuals were examined, each was doing the best that it could.

Parents that laid five eggs did so because laying six eggs would have led to a worse outcome. Parents laying seven eggs did so because laying more, or fewer, would have led to worse outcomes. Each according to his or her ability was the rule adopted by parent great tits, and the abilities of individuals differed. This was established by experimentally altering clutch size by transferring eggs across nests. Parents that normally reared large clutches fared better than those that normally reared small clutches when challenged with experimentally enlarged broods. The female tits were laying clutches optimal for themselves, not for the population as a whole. Not all great tits are created equal. Rather some are more equal than others.

Fit or Fat?

Do human mothers individually optimize brood size? At first glance it would seem that anything other than singleton births would be maladaptive. About 1 birth in 200 results in identical or monozygotic twins, and anywhere between 1 in 250 to 1 in 20 results in fraternal or dizygotic twins. Triplets and higher-order multiples are exceedingly rare in natural pregnancies. Humans are clearly not designed for multiple births, as even raising twins is a struggle. Twins are born earlier and smaller than singletons, and suffer higher fetal and neonatal mortality and morbidity than singletons. Mothers of twins similarly suffer higher mortality and morbidity than mothers of singletons. Medical scientists have gone so far as to declare that twinning in humans is an "atavism," a "hazardous deviation from the human norm," and perhaps most eloquently that "the human species is not designed to carry more than a single fetus in utero with any degree of biologic grace." Evolutionary biologists, too, have queried whether twinning is adaptive. As recently as two hundred years ago mothers with twins in rural Finland suffered higher mortality, longer interbirth intervals, and lower infant survival than mothers of singleton babies. Overall, mothers of twins left no more surviving children than mothers without twins, suggesting that twinning was either selected against or selectively neutral. Conversely, mothers of twins in northern Germany enjoyed a clear benefit over mothers of singletons: twinning evidently conferred a selective advantage in this population. And today in western Europe, North America, and Japan, unless they are born very early, twins and their mothers enjoy outcomes not dramatically different from those of singleton births. Is twinning adaptive or not? Over most of human history, the answer is probably not. Why then does twinning persist?

Twinning represents a classic quantity-quality trade-off—while increasing the size of the current brood, twins bear an array of developmental costs and pose a significant physiological challenge for their mothers. Is twinning adaptive, or might it represent a developmental accident or sequelae of other adaptive trade-offs? Behavioral ecologist David Anderson suggests that elevated rates of dizygotic twinning in older mothers arise as a side consequence of an insurance strategy to compensate for an age-related increase in embryo loss. Twinning might, therefore, represent a premium paid for insurance.

If twinning is a conditional strategy, then we would expect to find a correlation between mothers' prepregnancy condition and twinning frequency. And indeed we do. Mothers better equipped to meet the physiological demands of twins—mothers who are taller and fatter—are more prone to dizygotic twinning. And dizygotic but not monozygotic twinning occurs less often in malnourished women, which would further suggest a conditional reproductive strategy. Thus it appears that dizygotic twinning in humans is at least in part conditional on a mother being equipped to meet the burden of twins.

How does being heavier leave a mother better prepared for the twins? There are two obvious advantages. First, being heavier before pregnancy reduces the risk of underweight babies. Mothers with twins live closer to the energetic edge, and we can see this clearly in the relationship between maternal and nutritional factors and birth outcome. With singleton babies, even underweight mothers are likely to give birth to normal-size babies. Not so with mothers of twins. Unless an underweight mom with twins has access to enough food during pregnancy to allow her to add a lot of weight, she is likely to give birth to underweight babies. And even in mothers of average weight, the weight gain during pregnancy affects the size of their babies at birth: more weight gain means bigger babies. But such is not the case with heavy mothers. Whether they gain a little or a lot of weight during pregnancy has little effect on the weight of their babies at birth. Fat moms give birth to big babies. Being fat also means not having to worry about food shortfalls during pregnancy even if you are carrying twins. Thin moms enjoy no such luxury. Their twins will be born underweight if food becomes short during pregnancy.

The second advantage to being fat is that such a mother is better prepared to meet the heavy nutritional demands of twins following birth, and two most definitely cannot be fed as cheaply as one. The fat stores laid down during pregnancy are later converted to milk when the children are suckled. Even malnourished mothers can provide sufficient milk at first, but mothers with low fat reserves will run out sooner. The consequences are potentially disastrous for both mother and her babies. Greater fat stores provide a greater energy reserve for mothers to draw upon, and avert the risk of prematurely depleted milk supplies. Of course, being fat would have been far more important in premodern societies, and is still important in rural societies over much of the world today compared with first-world societies, where mothers enjoy the luxury of food available in virtually unlimited quantities.

A Womb for Two

Although it is obvious how being fat can help meet the burden of twinning, how might being tall play a role? Here it is a question of a womb for two. Twins have reached the same volume in the uterus by the end of the second trimester of pregnancy as a normal single-ton baby at term. From that point on, space becomes increasingly tight. Twins grow as rapidly as singletons over the first thirty weeks of pregnancy, but after this twins grow slower. By this time the pla-centas are likely to abut one another, and womb mates compete for both food and space. The advantage of being tall? More room in the womb, the difference between a bachelor suite and a one-bedroom apartment. As a consequence, taller women enjoy better pregnancy outcomes from twins than shorter women: their babies are larger and less prone to developmental problems.

Natural Selection on Twinning Frequency

I argue here that twins are maladaptive but that a strategy of twin-ning is not—its adaptive value lies in providing a hedge against de-fective embryos. Twin births are a cost, contretemps to be avoided. When and where this cost is relaxed, as in older mothers with lower rates of embryo survival, the frequency of twin conceptions rises. But the costs may also reflect geographic circumstances. One exam-ple that has been studied is the archipelago of Åland and Åboland in southwestern Finland. Historically, twins have been unusually common in this population. The question is, why? Unlike most human populations, in which twin births usually result in fewer sur-viving offspring when measured over the entire reproductive life-time of a woman, the penalty for having twins in this insular popu-lation is, or was from the mid-eighteenth to mid-nineteenth centuries, relaxed. Mothers of twins enjoyed roughly the same life-time reproductive success as mothers of singletons only, unlike the outcome on the adjacent mainland, where mothers of singletons enjoyed a clear advantage. And when only unlike-sex twins were examined, restricting the analysis to dizygotic twins, mothers of twins on the islands enjoyed a reproductive advantage over mothers of singletons. This step is important, as only dizygotic twinning is heritable (with rare exceptions) and thus subject to natural selection. The authors of the study, Virpi Lummaa, Erkki Haukioja, Ristor Lemmetyinen, and Mirja Pikkola, argue that predictable food from farming and fishing on the archipelago reduces the variance in reproductive success and renders twinning more affordable. This is an intriguing variant of the bet-hedging argument (see chapter 3). Twins would normally represent a risky high-variance reproductive option, as they are more vulnerable to resource shortfalls than singletons, and thus would be penalized for this relative to singletons. But if this variance is diminished by greater environmental predictability, as on the archipelago, the gamble may be worth taking.

A further curious twist to this story is that the success rate of twins was closely linked to their sex. In both the mainland and insular populations, twin daughters fared better than twins of unlike sex or twin males, likely reflecting a greater cost for sons during gestation than daughters.

Brood Reduction before Birth?

The notion that mothers might use prenatal brood reduction as part of an adaptive strategy of brood size management is one that deserves closer scrutiny. One is tempted to suggest that it makes too much sense for human mothers not to implement it. Statistically, humans are obligate brood reducers. According to Charles Boklage of the Department of Pediatrics at East Carolina University, perhaps as many as one in eight pregnancies begin as twins. But for every twin birth that follows, there are ten to twelve twin pregnancies that result in singleton births, and perhaps as few as one in fifty twin conceptions survive intact. These estimates have been challenged by Australian biologists studying the early stages of pregnancy using ultrasound, and their work has been widely cited in the popular press as debunking the vanishing-twin theory. But they examined only clinically recognized pregnancies and bypassed the critical early period around the time of implantation, when most losses occur. Also, their results do not square with survival rates reported from many studies of assisted reproduction (which represent an experimental manipulation of brood size). And the number of twin pregnancies examined was far too small for such a bold claim to be taken seriously; the results need to be replicated with a sample at least an order of magnitude larger.

Obligate brood reduction occurs in taxa ranging from sharks to pandas to pines but is perhaps best known in large predatory birds. These include tropical boobies and eagles, temperate pelicans and cranes, and penguins that are just plain cold. All these birds typically lay two eggs but raise only a single nestling. Usually. In every species that is described as an obligate brood reducer, exceptions to the rule are reported. Two offspring survive, albeit rarely. Whether this is accidental or deliberate is unclear, but to resolve the problem of definition ornithologists have adopted the 90% rule. At the population level, brood reduction is obligate if at least 90% of doubleton clutches or broods are reduced to singleton offspring. And brood reduction indeed occurs in more than 90% of twin conceptions in humans.

This high rate of attrition of twin conceptions is the vanishing-twin syndrome. Many twin pairs are trimmed to a singleton fetus, usually in the first trimester. And many more result in failed pregnancies. We as a species, therefore, qualify as obligate brood reducers alongside black eagles, brown boobies, and white pelicans. Exactly how it occurs in humans is unclear, but infanticide is a routine component of obligate brood reduction everywhere else, by parental neglect as in cranes, grebes, and pandas; by sibling cannibalism as in sand tiger sharks; or by fatal sibling competition as in eagles, pelicans, and penguins. That is what I shall turn to next.

Chapter 9

Fatal Sibling Rivalry

Siblicide

The Roman Empire had endured a succession of tyrants and despots when the emperor Pertinax reluctantly came to power. He set out to right the wrongs of his predecessors, to purge a corrupt administration, and to lift the burden of oppression from Romans. For this the Praetorian Guard, who had grown accustomed to the perks of power, quite naturally murdered him. These second-century thugs then proceeded to auction the emperor's throne to the highest bidder. Didius Julianus, a wealthy senator, outbid another pretender and claimed the throne albeit briefly. Julian's reign ended when the general of the Pannonian armies, Septimus Severus, deposed him. Severus in turn reverted to form and ruled as a cruel tyrant. He had two sons and potential heirs by the empress Julia: Caracalla and Geta. Almost from birth the brothers were fierce rivals. This troubled Severus, and though normally the elder Caracalla would have enjoyed the right of primogeniture, both brothers were anointed as successors to the emperor. Upon their father's death, the two were proclaimed emperors of Rome and ruled with equal and independent power.

Caracalla and Geta never reconciled, and an extraordinary bargain was brokered. The Roman Empire was to be split asunder. Geta would rule the east (Asia and Egypt), and Caracalla would rule the west (Europe and Africa). But before this unpopular and potentially disastrous deal was concluded, Caracalla opted for the simpler solution of siblicide. He murdered Geta, who died in the arms of his mother. Caracalla then ruled as a cruel tyrant until he too was murdered by the Praetorian Guard. The emperor Severus was no student of sibling rivalry. By opposing tradition and anointing his sons as coequals, he did not achieve his goal of eliminating conflict but rather quite the opposite. He had sown the seeds of maximum instability, and siblicide was the final result.

Why should sib murder sib, or parent kill offspring? Infanticide is not an obvious route to enhancing Darwinian fitness, yet is widespread among both plants and animals: burying beetles eat their own babies; nestling eagles and pelicans bludgeon younger siblings into an early death; and even tropical trees poison seeds that share the same pod. We expect warm and loving relationships among family members, and for this reason the study of infanticide was long swept under the academic carpet. It gained respectability only when we had a reason to look for it, and that reason was kin selection.

The logic of kin selection can be tidily summarized in a simple inequality:

$$Br - C > 0,$$

where C is the cost to self of a given act, B is the benefit to a relative, and r is the coefficient of relationship. Known as Hamilton's rule for its author, the late British evolutionary biologist W. D. Hamilton, it says this: an individual should act unselfishly if the benefit to the relative weighted by the degree of relatedness exceeds the cost to self. This is enlightened genetic self-interest, and C is what economists refer to as the opportunity cost of a given behavior. What could you have done with the same resources had you taken a different decision? Hamilton's rule is of course derived from Hamilton's theory of kin selection, and provides a rough but useful sketch of how family members should behave toward one another.

A simple example might be useful. Imagine that two full sibs ($r = $ ½), alpha and omega, share a nest. Alpha is older and stronger than omega, and can effectively decide how food delivered by parents is shared. When should alpha take the food for itself, and when should it give it to omega? Hamilton's rule provides the answer for us. The food item should go to omega if omega's fitness benefit (e.g., an increased probability of survival) is at least twice that alpha would obtain if it ate the same morsel of food. Imagine that the food item would increase omega's fitness by 10 units, which after being weighted by the coefficient of relatedness yields a net benefit of 5 units (10 units × ½). If alpha is already well fed, its fitness benefit for consuming the same food morsel might be quite small, say only 3 fitness units. Since $10 \times ½ - 3 > 0$, unselfish behavior is favored. The food item should go to omega. But now imagine that alpha has not fed recently and is somewhat undernourished. Under such circumstances its fitness benefit would grow to perhaps 7 units instead of 3. In this case Hamilton's rule prescribes selfishness, since $10 \times ½ - 7 < 0$. Alpha should take the morsel for itself.

Here the coefficient of relatedness, r, is key. It is the probability that two relatives will share a gene by common descent. If alpha and omega were unrelated ($r = 0$), selfishness would always be favored. This explains the often cruel behavior of brood parasites (see chapter 6). Even if alpha and omega were half sibs with $r = ¼$, Hamilton's rule would still predict selfishness. I'll leave it to you to work out the math, but the point is this. Success in the evolutionary game is measured not only in the success of one's own genes but in the success of identical copies, which are most often found in close relatives—a point Richard Dawkins made long ago with disarming clarity. Hamilton's rule sets the limits for both selfish and unselfish (altruistic) behavior, and provides us with a tool for understanding the often puzzling relations among siblings, whose brotherly love sometimes takes a sinister turn.

Hatching last in a black eagle or white pelican nest is a death sentence. The doomed offspring gains a reprieve only if its older and stronger sibling—its designated executioner—dies first. And the odds of that happening are long. The life of a last-hatched chick is usually poor, nasty, brutish, and decidedly short. Such obligate siblicide, one form of obligate brood reduction (see chapter 2), occurs frequently in large and long-lived predatory birds. The single offspring that survives from the initial clutch of two grows slowly over a period of dependence that lasts many months and may extend well beyond the time that chicks leave their nests.

Why don't parents raise more babies? By rearing the chick from the second egg they could potentially double their reproductive output. Surely that would pay—but perhaps not. These birds live long lives, and the rates of adult survival are exceptionally high: 90–95% of adults alive now, sometimes more, will be alive at the same time next year. With a 95% survival rate, an adult can expect on average to live twenty years. Here small changes have big effects. A slight drop in survival, from 95% to 90%, halves the expected life span, from twenty to ten years. And rearing more babies adds to the cost of parenthood. The extra work may reduce general body condition and the level of fat stores that parents need to tide them over temporary food shortfalls. It may interrupt the normal cycle of feather replacement and impair flight efficiency. Or it may suppress the immune system and leave individuals more vulnerable to parasites or pathogens. Or parents that need to make more foraging trips to bring more food to an enlarged brood may face additional risks from predators, a shark or seal ready to snatch a diving seabird. Thus doubling one's reproductive output in any given year may not be a bargain at all. There may be good reason for a seemingly leisurely approach to parenthood.

Parent black eagles, brown boobies, and white pelicans stand aside as alpha murders omega, but their role is one of more than just indifference. They are complicit in this foul deed as they handicap omega by hatching their brood asynchronously. This ensures a one-sided battle with the outcome virtually automatic unless disease or defect intervenes to weaken alpha. Obligately siblicidal birds practice a grim form of avian primogeniture. The hatching intervals are unusually long for birds, typically three to seven days, sometimes more. This ensures that the ensuing battle is lopsided and brief, and that the winner will emerge relatively unscathed. But what if the sibs are evenly matched, as was the result of Emperor Severus's ill-fated decision. Equal competitors engage in prolonged, bloody battles that benefit no one, hence the big advantage (earlier hatching, primogeniture) that is usually conferred on one of the progeny. Such handicaps, if equalized by a fair-minded parent, would undermine any progeny choice benefits (see chapter 7) of sib competition, as even a weakling given sufficient advantage can defeat a stronger opponent. Even I might have a chance of upsetting Gary Kasparov in a chess match if we took away his queen, two rooks, and a knight (OK, both knights and a bishop as well). There is a dark parental logic to ensuring one-sided battles. If a senior sib cannot defeat its younger rival given the commanding advantage bequeathed by its parents, then it probably was not worth rearing anyway.

Desperado Siblings Result from Extreme Favoritism

Sibling violence that occurs in the avian family Sulidae—the boobies—is both spectacular and brutal. The elder progeny in obligately siblicidal masked and brown boobies are merciless executioners. At most nests in most locations, the death of the younger sibling occurs unconditionally, with a reprieve granted only when the first-laid egg fails to hatch or the first-hatched chick is burdened with a congenital defect. In facultatively siblicidal blue-footed boobies, the aggression is more measured, conditioned in part by prevailing levels of food and in part by the establishment of stable dominance relationships.

Why are some species obligately siblicidal? We are still not sure. Perhaps at an ultimate level such behavior stems from scarce resources, though data bearing on this are scant. To be sure, the prospects of the junior sibling in brown booby nests are bleak. What is an individual who is the designated victim of siblicide to do? Meekly submit to a near certain death, or resist with the faint hope that its older sibling is too feeble to act as executioner? Under the desperado sibling hypothesis, if the younger sibling has nothing to lose, and if resistance does no or little harm to an elder sibling who shares the same gene, and if there is a nonzero chance of achieving a dominance reversal against a much older and larger sibling, then escalated aggression just might pay.

Our knowledge of the behavioral strategy of the victims in this bloody game is slim. One of the problems is that most of the younger sibs are eliminated so quickly that there is little opportunity to observe their behavior. Are the younger sibs as aggressive as their older brood mates? Hugh Drummond and coworkers employed a clever strategy to gather a behavioral profile of the junior brown booby sibs: they reared them in broods of a closely related species, blue-footed boobies. These results were then contrasted with the behavior of subordinate bluefoots fostered in bluefoot nests and unmanipulated chicks in both blue-footed and brown booby broods. The results were straightforward: obligate brood-reducing brown booby chicks were far more aggressive than facultative brood-reducing bluefoots in the same roles, so much so that they sometimes intimidated the much older, larger, and stronger blue-footed brood mates. And in almost half of the experimental broods, the brown booby juniors were relentlessly aggressive.

This of course came as a shock to the senior blue-footed booby chicks. Normally their younger nest mates accept their subordinate role and, by the avian equivalent of a gentleman's agreement, live more or less amicably with their brood mate. Should food be sufficiently short to suppress the elder chick's growth, however, the aggression turns lethal. Otherwise the aggression remains at sublethal levels.

Brown booby chicks are different. They escalate the fighting, usually in a losing effort. Drummond and coworkers argue that the difference in aggressive behavior of the marginal browns and bluefoots is driven by the differing prospects of subordinate chicks in blue-footed and brown booby nests. In short, the younger browns have nothing to lose by being aggressive, and something to gain. If they stand by meekly, even a feeble senior sib will defeat and kill them. But if they resist, they may occasionally deter their otherwise certain execution. The faint hope of a coup d'état maintains extreme violence in these extraordinary birds.

The Good and the Best

Closely related to brown boobies, and also obligately siblicidal, are the Nazca boobies found in the Galapagos. They are also closely related to the gannets found in the North Atlantic and elsewhere, and look a bit like gulls on steroids. In a clever experiment, behavioral ecologist Dave Anderson and his students tricked Nazca booby parents into raising doubleton broods, instead of the normal single chick from a two-egg clutch. The parent boobies were quite capable of rearing the larger than normal brood but paid a heavy cost to do so. The female parents' mortality rate more than trebled in the half year following the experiment (male parents were not as strongly affected). Just as important, the babies did not fare well. Their frequency of survival to breeding age was only half that for nestlings in singleton broods. Experiments on other seabirds have shown similar effects on offspring survival. Though two offspring may survive to leave the nest, they are almost always malnourished relative to singletons, rendering them less likely to survive the risky period of avian adolescence. Bigger is indeed better. In Nazca boobies the tandem of elevated maternal mortality and reduced survival of offspring means that parents with one robust chick seem to do better than those with two smaller progeny: obligate siblicide pays off.

But there may be more to the story than just the simple survival of parents and, in particular, offspring. Long-term studies of bird populations in which the fates of individuals are followed from year to year reveal something very interesting. A minority of parents produce the majority of all progeny. Given the competitive breeding environments faced by obligate brood-reducing species, it is probable that not all offspring reaching independence enjoy an equal chance of becoming successful parents. More likely it is the premium-quality offspring, which are best able to compete for food, space, and mates, that will become successful parents. As Voltaire said, the best is the enemy of the good. Unfortunately, the key data are missing. One needs to gather data on lifetime reproductive success to test this idea, and that is no easy task for species whose members may outlive the biologist.

There is an obvious question not yet addressed. If parents are so manipulative in reducing the size of their broods, why should their offspring play along with the scheming? Why should sib murder sib? The answer is largely the same as that for the parents. Resources are finite, and one high quality progeny (Voltaire's "best") is better than two of lesser quality (the "good"), particularly when parents must mortgage their future to pay the cost of rearing extra offspring. But the decision for offspring includes another key dimension: evolutionary narcissism. Whereas parents are equally related to all their progeny, alpha is, in effect, more related to itself than to omega, unless alpha and omega happen to be identical twins. If the parents are conservative in their reproductive decisions, then their offspring are even more so. They are ultraconservative: if parents favor obligate brood reduction, it is even more certain that alpha will too.

Facultative Versus Obligate Brood Reduction

Bald eagles practice facultative brood reduction: brood reduction if necessary, but not necessarily brood reduction. And parents are sensitive to the unfolding rivalries, though they face an added challenge. The offspring are sexually dimorphic, females being 20–25% larger than males. If a female hatches after a male nestling, it will eventually close the size gap and overtake its brother, leading to fierce battles if food happens to be short. Mother bald eagles, though, seem to have found a way around this dilemma. When food is likely to be short, they hatch their female chicks first in mixed-sex broods. Male-first broods are rare. This ensures that a stable dominance hierarchy exists and diminishes the risk of costly conflict. But context is everything for bald eagles: nestlings that engage in murderous battles when food is short show a different personality when food is plentiful. Older nestlings may even tear apart prey to feed younger siblings. The distance between conflict and cooperation is often not great.

Hyenas also practice facultative siblicide but with an interesting twist. When food is short, unlike-sex twins are more likely to survive, whereas all-male and all-female broods are usually reduced to singletons. Unlike birds, hyena pups are born synchronously, and same-sex twins are like the sons of Severus: evenly matched, leading to escalated competition and perhaps otherwise unnecessary siblicide. But the outcome is conditional on food levels: when food is abundant, same-sex twins often survive. Again we see an avoidable conflict.

Food also plays a key role in facultative siblicide in birds. In ospreys and blue-footed boobies, aggression is intimately linked to hunger. Well-fed chicks live amiably, and hungry chicks fight. The bluefoots, though, adopt a two-track strategy. As long as the elder chick, alpha, in a doubleton brood grows at close to its maximum rate, the link between hunger and aggression remains, but its aggression is nonlethal. If, however, the growth rate of alpha falls below a threshold, roughly 75% of the maximum, a developmental switch is flipped, and it turns obligately siblicidal. In this system facultative aggression linked to hunger seems to serve as a hedge against short-term food deprivation.

In baby egrets the link between food and aggression is less clear. Douglas Mock and his students have shown that nestling cattle egrets and great egrets remain belligerent nest mates whether food is rich or poor, and it is the nutritional state of the last-hatched nestlings that is the chief determinant of their survival. A well-fed omega chick can withstand the continued beatings from its older nest mates. A malnourished nestling quickly succumbs.

In obligately siblicidal birds, the execution of omega is swift, ruthless, and obligatory. Curiously, this aggression is not about food, as siblicide occurs among black eagles even when the nestlings sit in a veritable sea of food. It is instead about opportunity. Alpha enjoys its maximum advantage over omega at hatching, and if food is likely to become limiting later on, when nestlings are eight weeks instead of eight days old, then there is no point in waiting, and from alpha's perspective there may be a heavy cost. Omega is surely going to become more difficult to kill as it grows older and larger, and may even overthrow alpha in the dominance hierarchy. The question for alpha is, why wait? And usually it does not. But not the possibility of food limitation later on is prerequisite for obligate brood reduction. It may well be that in some years there is more than enough food for two, but alpha cannot know that, at least not yet. It just does not pay alpha to show clemency toward omega long enough to find out whether there will be sufficient food for both, and in the interim the roles of victim and executioner may be reversed. By killing omega early, alpha protects its birthright, as did Caracella when he murdered Geta in his mother's arms. Alpha secures uncontested access to all future parental care.

Future access to food is also at the root of obligate brood reduction in flesh flies of the family Sarcophagidae. These flies are both common and unusual. They resemble houseflies but are much, much bigger, and in many the larvae develop on the carcasses of dead animals. Others seek out skin sores of vertebrates, and one deposits its embryo—it is a live-bearing fly—in pitcher plants. The predatory larva is covered with a waxy cuticle to shield it from the plant's digestive enzymes and sits inverted at the water's surface, bottom up, breathing through spiracles on its posterior. It feeds on the hapless insects that drop into the plant's pitcher, and it does not like company. Place two larvae together, and they begin a death struggle that ends when one submerges and drowns the other. Here the victor protects its feeding site and again secures sole access to future food.

Obligate brood reduction is surprisingly common throughout nature. Go to the nearest window and look outdoors. If there is an evergreen tree, it is likely an example of obligate brood reduction. Conifers routinely create two or more embryos for each seed, but only one ends up in the finished product; the second embryo provides a cheap form of insurance against a developmental defect, parasite, or pathogen. Sand tiger sharks play the same games. Mothers have twin uterine horns and deposit several fertilized embryos in each. The first to hatch hunts through the soupy fluid on a mission to search and destroy. Its prey? Its own brothers and sisters. Each uterine horn has room enough for only a single pup, which will eventually grow to a meter in length on a diet of unfertilized trophic eggs that mother provides (see chapter 11). Siblicide ensures that the firstborn shark will be the lone survivor when this game of internecine strife is finished.

Thus relatedness alone is no guarantee of merciful behavior: r is only one component of Hamilton's rule. One must weigh the net costs and benefits as well, and since alpha and omega, who are close relatives, live in even closer proximity, they are also the closest of competitors for present and future resources. With siblicide, selfishness among close relatives approaches an extreme. Could it get any worse?

Ultraselfish Alleles

Without the aid of the sorceress Medea, Jason and the Argonauts would not have conquered the sleepless dragon and recovered the Golden Fleece. Nor would they have eluded Medea's father, the Colchian king Aeetes, for Medea had taken hostage her younger brother Absyrtus. When the vessels of Aeetes closed on the fleeing Argonauts, she killed and dismembered her brother and cast his limbs upon the sea. Jason and the Argonauts and their passenger Medea made their escape while Aeetes recovered the remains of his murdered son. But Medea's services came at a price, a promise of marriage. Jason and Medea were wed, had two children, and retired to Corinth, where they did not live happily ever after. Perhaps because she was foreign-born, or because she was getting on in years, or just because she was a bloodthirsty sorceress, Jason divorced Medea after a decade of marriage in favor of the younger princess Creusa. Jason should have known better. Medea was not a woman to be scorned. In a jealous rage she poisoned Creusa and then murdered her own children before fleeing to Athens in a serpent-drawn chariot. You just knew it was not going to end well.

Medea, who murdered repeatedly to achieve her goals, lends her name to an equally sinister form of selfishness among siblings. Imagine a gene for siblicide. Normally, this would be a gross oversimplification, as most traits are the product of many genes working in concert, but in this case the assumption may not be so far-fetched. If such a siblicide gene harmed only nonbearers, what would stop it from spreading? We have to take a gene's-eye perspective to follow the logic. Imagine further that this gene could recognize who had the gene and who did not. As we expect genes to look out for themselves and for identical copies of themselves, the rule might be this: clemency for other individuals who bear this gene and execution for those who do not. What if you had this gene but your brother did not? Relatedness is important only insofar as it provides a guide to the likelihood of sharing copies of identical genes by common descent. The coefficient of relatedness, r, is this probability. But if you can replace that probability with a certainty—a given individual either does or does not have the same version of a gene as you—then r becomes unimportant. Flour beetles, *Tribolium*, have a secret. They know.

The system is simple and chilling. It is a single version of one gene, called the MEDEA allele. In this case MEDEA is an acronym for Maternal Effect Dominance Embryonic Arrest. The MEDEA or M allele does this: it kills all sibs without it and leaves sibs with it unharmed, and seems to work by producing both a poison and antidote. Bearers of the allele are thus protected from the effect of the poison; nonbearers are not.

Until now I have been rather loose in my terminology for genes. All members of a species share the same complement of genes, just as all baseball teams have a similar roster of players: a pitcher, a catcher, a center fielder, a shortstop, and so on. But teams differ in that they have different players at the same positions. So it is with a roster of genes. All members of a species have the same gene "positions," but the players at those positions can differ. Throughout a population, there are usually multiple versions—alleles—of any given gene. Individuals in sexually reproducing species have two copies of every gene, and the two alleles can be the same or different. If they are the same, they are referred to as homozygous; if they are different, then they are heterozygous. New alleles arise as mutations of existing genes, and when they enter a population, they are necessarily rare. In sexually-reproducing species, the first mutant allele will always occur in the heterozygous condition.

Now imagine that a flour beetle mother carries a copy of the M allele. We shall call the other version of the same gene the wild-type allele, denoted +. The mother's genotype will thus be $M+$. She mates with a male who has two copies of the wild-type allele. His genotype is ++. In sexually reproducing species, every gamete (egg or sperm) will receive one of the two alleles. For mom, half her eggs will have the M allele, the other half will have the + allele. For the male, all his sperm will have the + allele. When sperm and egg join in fertilization, there will be two possible genotypes in the offspring. Half will be $M+$, and half will be ++. This is not good for the ++ offspring.

They are certain to die, because they do not have *M* allele. Only the *M*+ offspring survive. In a clutch of *Tribolium* eggs, half of the off-spring are doomed.

Consider this from the mother's perspective. She carries inside her a ticking time bomb (*M*) that will ultimately destroy half of her off-spring. She loses. The winner? It is the *M* allele. It ensures that all surviving offspring carry it into the next generation. The fix is in, and this allele will quickly sweep through the population and elimi-nate the + allele entirely.

There are probably many such MEDEA-type alleles in many organ-isms. We simply do not see them, because they act quickly and ruth-lessly. The *M* allele was not discovered in flour beetles until two geo-graphically distinct strains were crossed. A quarter century ago, Richard Dawkins, building on Hamilton's work, proposed his selfish gene theory. It was then steeped in controversy, and there are still those who grumble about it today. But for evolutionary biologists, it is now an essential conceptual tool. We cannot explain MEDEA alleles and an array of related traits without selfish genes.

Obligate brood reduction spans the taxonomic spectrum, from beetles to trees to birds, but interestingly it appears to be unusual in mammals. It just may be that brood reduction in mammals occurs out of sight: twin pronghorn embryos are relatively common in utero, for example, but rare at birth. Siblicide here appears to occur in the womb, with one embryo growing a spear of tissue into the other. This prenatal brood reduction yields a benefit of reduced post-natal competition for mother's milk.

Human Twins

Such internecine strife occurs closer to home than you might think. Twin births are relatively rare in humans, though surprisingly com-mon at conception (see chapter 8). Ultrasonography, which became widespread in the 1970s and 80s, led to the detection of this van-ishing-twin syndrome. We now know that a remarkably high pro-portion of twin conceptions undergo early brood reduction, re-sulting in the birth of a single baby or none at all.

About two-thirds of twins born are dizygotic (DZ) (see chapter 8 for the difference between monozygotic and dizygotic twins). They arise from a pair of fertilized eggs and are no more closely related than any pair of siblings. Indeed, these are often referred to as frater-nal twins. The remaining one-third of twins born are monozygotic

(MZ), arising from the fission of a single fertilized egg, and they are genetically identical—in effect, natural-born clones.

Monozygotic twins differ according to their early anatomy. Here I need to describe the early stages of pregnancy. The human placenta is a disk-shaped structure that grows on the uterine wall to the size of a dinner plate. It is a coproduction of mother and of the embryo's chorion, a membrane that surrounds the fertilized egg. The placenta is the life-support system for the embryo, and is where the blood systems of the mother and embryo, though separate, come into close contact. The contact is sufficiently close that the embryo can draw life-sustaining nutrients and oxygen. The embryo grows out from the placenta, connected to it by the umbilical cord, and it grows inside a fluid-filled sac, the amniotic cavity. The rupture of this sac (water breaking) occurs at the end of gestation and signals the onset of labor.

An egg may split anytime in the first two weeks after fertilization, resulting in monozygotic twins. The timing of the split determines which of three types of MZ twin occurs, and they are classified according to the structure of their chorion and amnion (the membrane surrounding the amniotic sac). The twins can be dichorionic and diamniotic: two separate placentas and two separate fluid-filled sacs. Or they can be monochorionic and diamniotic: the placental disks of the two embryos are fused together into a single unit, but there are still two fluid-filled sacs. Or they can be monochorionic and monoamniotic: one fused placenta and one fluid-filled sac.

The type of monozygotic twins that arise depends on the timing of the split of the egg following fertilization. If it divides in the first three days postfertilization, dichorionic-diamniotic twins arise. Here the separation of the embryos occurs sufficiently early for the two twins to grow independently of one another. If the egg splits four to eight days after fertilization, monochorionic-diamniotic twins arise. They now share a fused placenta but grow in separate amniotic sacs. If the split occurs from day nine to twelve, monochorionic-monoamniotic twins arise: the embryos share both a common placenta and a common amniotic sac. These are very close relatives, physically as well as genetically. But even closer relatives arise if the embryo divides very late, thirteen or fourteen days after fertilization. This gives rise to conjoined or "Siamese" twins, an exceedingly rare event occurring in only one in fifty-thousand pregnancies.

Close contact before birth is dangerous for human twins. Dichorionic twins grow independently of one another (all DZ twins are of this type), and their prospects are comparatively good. But monochorionic twins are forced by an accident of circumstance into an

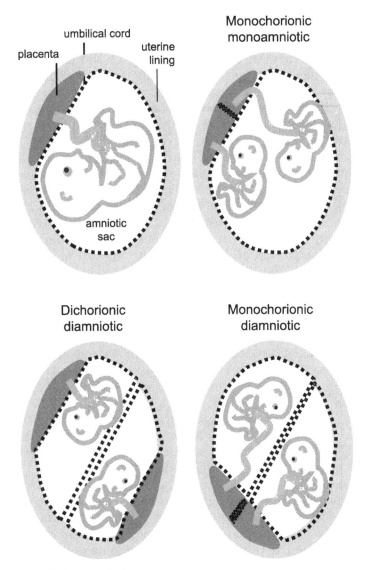

Figure 9.1. Twin types in humans
The structure of the placenta and amniotic sac varies in different types of human twins. The top left panel shows the location of the placenta, amniotic sac, uterine lining (endometrium), and umbilical cord in a singleton pregnancy. All dizygotic twins (from two eggs) and some monozygotic twins (from one egg) have a separate placenta and amniotic sac (*bottom left panel*). Some monozygotic twins share a fused placenta but have separate amniotic sacs (*bottom right panel*). Still other monozygotic twins share a fused placenta and common amniotic sac (*top right panel*).

internecine and often fatal struggle. Like Geta and Caracalla, they are too evenly matched, and one or both are likely to perish. Even the fact that they are genetically identical provides no shelter from this conflict. They are forced to share space and maternal sustenance, but they share much too much, and both risk ending up with nothing at all.

The technical term for what ensues is twin transfusion syndrome. Monochorionic twins share a common blood supply by vascular linkages that connect their two halves of the placenta. This should not be. The fate of the twins now rests on maintaining an exact and precarious balance. If this equilibrium is maintained, all may turn out well. But if the equilibrium is perturbed, the results are often fatal. It begins when the blood flow of one twin, the "donor," begins to overpower that of the other, the "recipient." The blood volume of the recipient grows and that of the donor falls, with inimical effects for both. The growth of the donor twin, deprived of oxygen and nutrients, slows and lags behind that of the recipient. This is one reason why monozygotic twins are often far from identical. Indeed, donor and recipient twins are easily differentiated at birth, the recipient appearing pale alongside the donor. That Esau came out of the womb red, and more vigorous than Jacob, hints that the sons of Isaac and Rebekah may have been the first donor and recipient twins.

The excess fluid volume is also potentially damaging to the recipient. Again timing is everything. If the reversal of blood flow occurs early, before organ formation, the recipient may fail to grow a heart, a brain, kidneys, or a pancreas. The donor becomes the "pump" twin, as its heart must sustain blood flow for itself and its doomed twin. Here the death of the recipient is certain, and the prospects for the pump twin are not much better.

Congestive heart failure is normally associated with the elderly. It arises when a damaged heart becomes unable to pump excess fluid; as a result the heart swells and grows ever more feeble until it gives out entirely. This customarily follows a prolonged bout of heart disease, but congestive heart failure may occur in utero. If the reversal of blood flow between MZ twins described above occurs after the organs have been formed, the excess fluid may overwhelm the circulatory system of the recipient. As with Jason and Medea, this is sure not to end well. Recipients that survive until birth begin life with a damaged heart, twin transfusion syndrome being one cause of congenital heart defects. Most of these twins, however, perish before birth. Unlike siblicidal birds, for which the death of a nest mate yields a benefit to the survivor due to relaxed competition for food, no such benefit accrues to the surviving twin in twin transfusion syndrome. Its journey has in fact become even more treacherous.

The first peril is immediate. Throughout early development, the blood flow of each connected twin has pushed against the flow of the other, as though the two were on opposite sides of a door. When one twin dies, the door suddenly swings open, and the blood of the twin on the other side rushes out. The result is a swift and massive blood loss that may cause the sudden death of the second twin. Even if the second twin survives this immediate crisis, further danger still lurks from the debris that will flow through the vascular connection from the now dead twin. This may result in blood clotting that ultimately proves fatal to the (temporarily) surviving twin. Something similar may have caused the death of the famous "Siamese" twin Chang after his conjoined twin Eng died from pneumonia. Neurological damage is another potential danger, with cerebral palsy being especially common in the surviving twin.

The numbers suggest that monozygotic twinning in humans represents obligate brood reduction, though the exact figures are difficult to untangle because the earliest stages of pregnancy, when most embryonic mortality occurs, are still poorly known. In the bird world, the rule of thumb is that brood reduction is obligate when it occurs in 90% or more of all broods. Such appears to be the case in twin pregnancies in humans, and is almost certainly true for monozygotic twins. As I noted in chapter 8, perhaps one in eight conceptions involves twins, both dizygotic and monozygotic, but only one in one hundred births results in twins. One or both twins thus perish in better than nine-tenths of all twin conceptions, surpassing the 90% threshold. And the rate of brood reduction in monozygotic twins is far higher than that in dizygotic twins. We can tell this by examining the relative proportions of monozygotic and dizygotic twins at birth and in embryos that are spontaneously aborted by mothers. Though dizygotic twins outnumber monozygotic twins two to one at birth, MZ twins outnumber DZ twins six- to eighteenfold in spontaneous abortions. Evidently only a tiny proportion of monozygotic twin conceptions survive to birth.

Why, then, obligate brood reduction in humans? Most mothers are not designed to carry two babies, and here an interesting difference between monozygotic and dizygotic twins emerges. Monozygotic twinning is quite clearly maladaptive for humans. The complications that arise during pregnancy most assuredly render it unprofitable, as most MZ twins do not survive until birth. Unlike DZ twinning it is not heritable or, if it is, only very, very weakly so. It also seems to be unaffected by prevailing environmental conditions, whereas dizygotic twins become less likely during periods of food scarcity. Thus, mothers in an evolutionary sense do not "choose" to have monozygotic twins. The rates of monozygotic twinning are remarkably constant: roughly 3 to 5 births per 1,000 pregnancies in all societies, whereas the rates of dizygotic twinning vary widely, from 1 in 20 births in Yoruba women to 1 in 200 for Japanese women. Dizygotic twins tend to be conceived by women capable of carrying twins when environments are favorable. Monozygotic twins appear to arise as accidents during early development whether mother is ready or not. She is forced make the best of a bad job, and though a mother may not have "chosen" to have monozygotic twins, she controls whether to carry MZ twins to term. Most often she does not, thus obligate brood reduction.

Dizygotic twinning, however, is both strongly heritable and subject to environmental influences. It is an evolutionary "choice" subject to natural selection. Before going any further I want to define clearly what I mean by evolutionary choice. I *do not* mean a conscious decision. I do mean that dizygotic twinning reflects the mother's ability to pay the heavy burden of bearing twins, both before and after birth, and is part of an adaptive strategy crafted by natural selection. Thus, tall mothers with room for two, and plump mothers with ample nutritional reserves, and mothers in favorable ecological circumstances are those most likely to bear dizygotic twins. Human mothers "choose" to have dizygotic but not monozygotic twins, which helps to explain why they are far more likely to carry a DZ twin conception to term. Humans thus share a trait in common with beetles, birds, flesh flies, and trees. We are all obligate brood reducers.

In this chapter we have explored a dark netherworld of family relations, a bleak landscape with indifferent parents and mercenary progeny, of selfish and even ultraselfish genetic elements. But under the light of Hamilton's rule, it all makes sense in a cold evolutionary calculus.

"Biological" Influences

Blue-footed boobies and ospreys, among which sibling aggression is clearly food sensitive, provide an arresting illustration of the intricate relationship between genes and environment. For some curious reason, the adjective *biological* has taken on the colloquial meaning "hard-wired behavior with a genetic basis" or, simply, "genetic determinism." Professional students of behavior, such as myself, find this last term wrongheaded and offensive, and equating *biological* with "genetic determinism" is a slander against my chosen profession of biology.

Behavior lies at the interface between the genetic programming of an individual and the environment in which it finds itself. Some behaviors are loosely coupled to environmental influences, whereas others are more strongly linked. But the fact that an environmental influence can be demonstrated in no way lessens the import of the underlying programming. The behavior of blue-footed boobies is conditional on prevailing food levels: once the older chick's growth rate falls below the 75% threshold, a genetic subroutine is invoked that says kill your younger sib. The environmental influence is profound, but the underlying genetic influence is equally profound. Here we have a conditional strategy in which a response to the environment is molded to natural selection. Over evolutionary time, blue-footed boobies with this program left more descendants than others with different programming (e.g., programs that said kill younger sibs when your growth rate falls to 80% of the maximum, or 60% of the maximum, or never kill younger sibs).

Equally arresting is the logic that all behavior is environmental in origin, the theme that animates the discipline of sociology. From Durkheim onward, sociology has been defined by a search for the antecedents of behavior in the prevailing social environment or, simply, environmental determinism. The role of genes is at best ignored and at worst denied. Oddly, some sociologists seem unable to accept that behavior can evolve in the same manner as a bird's wing or fish's tail. And for all but a minority of sociologists willing to explore beyond the four walls of their very small box, the term *sociobiology* is an object of scorn. Again a key term has been appropriated by others for misuse. Sociobiology to most sociologists seems to mean the role of genetic (biological) influences on human behavior. Some go further, equating sociobiology and its sister fields of evolutionary psychology and anthropology with quack "just so" explanations of human behavior, a clear example of academic propaganda. For biologists interested in human behavior, these epithets are not only wrong but also dangerous. To begin at the beginning, sociobiology traces its origins to the publication of E. O. Wilson's (1975) volume of the same name. Wilson's *Sociobiology* contained an encyclopedic review of the social behavior of organisms ranging from microbes to primates, and described a program for the study of social behavior founded on a synthesis of behavioral, evolutionary, and population biology. It was primarily about the social behavior of animals. One chapter in twenty-seven, just 6% of the text, dealt with human behavior. Yet that 6% defines sociobiology for its critics. And this impedes progress, as Wilson's program—to search for the antecedents of social behavior in animals and see if they hold in humans—is a logical approach. Can we understand sibling rivalry or infanticide in humans by studying analogous behaviors in other species? Unless we have been completely uncoupled from a 3.5-billion-year evolutionary legacy—unlike any other organism—the answer is an unequivocal yes! Environment matters. So too do genes, sometimes in astonishing ways.

Chapter 10

Family Harmony

Cooperation in Families

Eros, the god of passion in Greek mythology, was the son of Aphrodite. His mother was troubled that her son remained forever as a child, and vexed, she took her problem to Themis, the wise goddess of justice who provided the solution. Eros needed a brother, for love must be returned to grow. Aphrodite soon gave birth to Anterios (Returned Love), and as Themis predicted, Eros grew rapidly in both size and strength. Here a brother's love was a necessary synergy for Eros to thrive.

Conflict and cooperation are logical counterparts, and both are routine components of family relations. In recent years the spotlight has shone most brightly on the potential for conflict (see *The Evolution of Sibling Rivalry* by Douglas Mock and Geoffrey Parker for a masterful review), and selfishness is no longer counterintuitive to evolutionary biologists. Baby sand tiger sharks hunt sibs in utero; penguins and blackbirds squeeze sibs, and white pelicans and black eagles pummel younger nest mates into an early demise. Tropical trees resort to poison; pronghorns, spadefoot toads, bee-eaters, and piglets develop modified weapons to slash, spear, and even cannibalize brothers and sisters. Such acts are all easily explicable as the logical albeit dire outcome of sibling competition for scarce resources. These are hardly the "family values" that many would like humans to emulate. With selfishness now the norm, the other side of family relations—unselfish or altruistic behavior—becomes increasingly interesting. In many cases the mechanisms for cooperation are just as counterintuitive as were the initial examples of family conflicts.

Just as siblings can be fierce and lethal rivals, they can also be stout and even irreplaceable allies. Hamilton's rule ordains both darkness and light in family relations—what matters is context. As Mr. Micawber remarked to David Copperfield: "Annual income twenty pounds, annual expenditure nineteen and six, result happiness. Annual income twenty pounds, annual expenditure twenty pounds ought and six, result misery. " Sibling rivalry is the inevitable outcome of a dearth of parental resources met by an excess of desire.

Replace that shortfall with a surplus, and adversaries are now potential allies. Limits to critical resources—food, warmth, and shelter—can even be relaxed by the synergies of cooperation among relatives. Nestlings or pups huddle together to share warmth and, by reducing the costs of thermoregulation, reduce overall food demands as well. By hunting in groups, social carnivores can bring down more and larger prey. And when close relatives work together, divide the tasks, and specialize roles, the results are often spectacular.

The Arthur Dent Effect

The dark side of parental optimism is that it encourages conflict within the family over scarce resources. But it also creates opportunity. Knowing the moment of one's death can be liberating. Progeny in organisms ranging from coelacanths to pea aphids to honeybees exhibit this Arthur Dent effect. Dent, a central character in Douglas Adams's *Hitchhiker's Guide to the Galaxy*, learns that he is not to die before arriving at the planet Stavro Mueller Beta. Armed with this foreknowledge, he engages repeatedly in reckless behavior without consequence until he returns to Earth: he meets his end shortly after uttering the infamous line "If I'm here, we're safe," not discerning that Stavro Mueller Beta was not a planet but rather the name of the nightclub in which he was presently standing.

An offspring that is freed from the responsibility of preparing itself for an independent existence, with no need to sequester resources for its own future survival and reproduction, can itself engage in reckless behavior. These progeny can now serve other family members by dying.

Why Cooperation?

Cooperation occurs within and outside families for a simple reason: groups can be more efficient than individuals. The reasons are threefold: specialization, an enhanced work capacity, and the opportunity for social learning. Social groups allow for increased efficiency by division of labor and specialization. We see breathtaking examples in the social insects: individuals within an ant or bee colony—effectively gigantic extended families—are morphologically specialized for their specific and clearly defined roles. Soldiers are armed with weaponry to defend colony mates—oversize queens are egg-manufacturing machines, supported by workers who attend to the mundane household duties.

Parent birds and mammals adopt sex-specific roles—for example, female mammals lactate, males do not. Hawks and eagles exhibit "reverse sexual dimorphism" (RSD), with females larger, often substantially so, than males. RSD is associated with a clear division of parental roles. Female ospreys (large, fish-eating hawks) are roughly 20% larger than males, a moderate difference in body size that corresponds to a marked difference in parenting roles. Females are stay-at-home moms, incubating eggs or brood nestlings, tasks that males rarely perform. Instead the fathers are responsible for provisioning the family, capturing fish that they return at regular intervals to the nest. Here specialization works. For the first month after hatching, the chicks are in need of more or less constant care to protect them from predators and to keep them warm. The bigger females are better suited to fending off crows, ravens, and other hawks, whereas the males, because of their smaller body size, are more efficient hunters. Smaller birds use less energy when flying, lowering the costs of food capture and transport—compact male ospreys can deliver fish to the nest more cheaply than the larger females.

Cooperation also means that social groups can potentially expand the capacity of work beyond the simple sum of the efforts of individuals—the advantage of synergy. The range of tasks that can be tackled is expanded to the benefit of group members. Lions, wolves, and killer whales that hunt in packs can kill larger prey—zebras, moose, blue whales—than solitary individuals. By foraging in groups, pelicans can herd fish into clumps that are easily plundered. Families are automatic social groups, and the payoff for cooperation is greatly increased by shared genetic interests. Individuals can be assigned specific roles to aid the group. Owlflies, for example, put proto-offspring to work to the benefit of other family members in an unusual fashion. The adults lay clumps of eggs on plant stems, where they are vulnerable to ant predation. The solution? They build ant-proof ramparts with unfertilized eggs, structures called repagula, to protect the fertilized eggs.

Living in groups also enhances the potential for social learning. Siblings can act as tutors for one another. Osprey chicks from doubleton broods, for example, learn complex hunting skills faster than singleton offspring. Shark and eagle biologists have conjectured that the struggles of siblings in utero or at the nest battle-harden the surviving progeny for life after independence, though evidence for such effects is scant.

The Road to Cooperation

Close proximity encourages interaction among individuals, with cooperation one possible outcome, conflict another. The path to stable cooperative arrangements among nonrelatives requires mutual benefit. These can be either simultaneous (mutualism) or, more rarely, asynchronous (reciprocal altruism). Living in family groups obviously lowers barriers to cooperation by allowing for kin selection. Costs to personal fitness paid by one individual can be recouped in benefits conferred on the genes of relatives.

Examples of simple mutualism in social organisms are commonplace: one often sees cooperative hunting, defense, and breeding. Here group members enjoy a simultaneous benefit. Meerkats, or dwarf mongooses, live in tight-knit family groups and use cooperation effectively to thwart potential predators. Sentinels stand guard to warn foraging individuals of the approach of predators, and they not only sound the alarm in the presence of danger but use a sophisticated language to alert the others to the type of predator approaching (duck—it's a hawk; run—it's a snake).

Florida scrub jays engage in cooperative breeding, with sons and daughters from previous broods often remaining at home to help rear younger siblings. Kin selection probably greases the path to such cooperation but is only part of the story. The adult pairs live on permanent territories that are in limited supply. With breeding vacancies few, the progeny face limited options upon reaching independence—they can strike out on their own and most likely move into a marginal breeding habitat, or they can stay on the parents' territory and help with the household chores. Sons and daughters play a ghoulish game, awaiting the death of mom or dad to inherit the territory, or wait for another breeding vacancy to arise, and in the meantime assist their parents in helping to feed younger sibs. Cooperation can also arise if favors conferred now are later returned, though bona fide examples of such reciprocal altruism are relatively scarce, so much so, that any well-documented example is worthy of note. Blood-feeding vampire bats provide one such case.

Vampire bats live in more or less permanent social groups and must obtain a blood meal every sixty hours or so or starve to death. Bats approaching starvation beg from recently fed bats, and the donor regurgitates a portion of its last meal to the bat in need, buying it an extra day or so to find another meal. An interesting physiological property of the system is that the cost to the donor is modest (it loses only a few hours on the starvation "clock"), much less than the benefit to the recipient. Bats that are recipients of such lifesaving acts are expected to return the favor at later dates, and indeed do, as every bat is prone to an occasional bad stretch.

What probably makes systems of reciprocal altruism relatively rare is the temptation to defect: to not return the favor, thus banking the benefit without paying any cost. For reciprocal altruism to work, the favors must be indeed reciprocal, requiring that cheaters be recognized and punished (there is no free lunch, or in this case blood meal). This by default means long-term social groups and individual recognition.

More often, among groups of relatives, kin selection provides the *appearance* of altruism, as when one individual, a parent or sibling, performs an act that is costly to itself to benefit a relative. In truth this is not a selfless act, as it is underpinned by genetic self-interest: a gene aids copies of itself that reside in the bodies of others. An obvious question is whether the remarkable behavior of vampire bats is just another example of kin selection. The answer appears to be no. Both relatives and nonrelatives live together, and both participate in the reciprocal exchanges. But families do provide the ideal forum for the evolution of cooperation: they have the essential prerequisites of tight-knit social groups bound together by a common genetic interest, and it is within the context of families that we see the most spectacular exhibitions of self-sacrificing behavior.

Parental Optimism and the Evolution of Cooperation

The benefits of cooperation reach their zenith when individuals become morphologically and/or behaviorally specialized for specific roles within a team. And specialization becomes more plausible with the production of surplus offspring under a strategy of parental optimism. Not all offspring, or even protooffspring, are destined for the path of reproduction. With some offspring on standby for a flight that may never arrive, the potential for adopting other roles grows, with the result that some progeny are fated for dead-end lives as food for others, as a sterile soldier caste, or even as fratricidal sisters. Redundancy is a key step in the evolution of new functionality. With marginal offspring often finding themselves surplus to their parents' needs, the opportunity arises to take on new functions under the third category of incentives for parental optimism: facilitation. But this third tine in the trident of parental optimism differs in an important way from the first two of extra reproduction and replacement: selection may favor traits that impair an individual's prospects for future reproduction. Under the extra reproduction and replacement incentives, individuals must be ready to go, just as a backup quarterback must always be ready if the starting quarterback falters. Under facilitation, marginal progeny may take on new roles that add value to their presence (such as becoming the football team's placekicker), roles that may even erode their value as a stand-in for or addition to the core brood (e.g., time spent practicing as a placekicker may reduce the time spent practicing as a quarterback and erode effectiveness in that role). Compromising an individual's reproductive future is a key step in the evolution of sibling cooperation: once this threshold is crossed, it opens the door to spectacular systems of self-sacrifice and teamwork.

The first step on the path to facilitation may be simple opportunism. The built-in redundancy of a strategy of parental optimism entails an element of waste: resources are invested in offspring that may not be needed. Recycling can reduce this waste. Infertile eggs and surplus or damaged progeny are cannibalized by parents and sibs, allowing other family members to reclaim otherwise squandered nutrients. And in some organisms, such opportunistic consumption of biological debris has quite clearly evolved into routine and even obligatory components of reproduction. Ants and bees use larvae as a food cache that can be drawn upon in times of need. Sand tiger sharks use unfertilized eggs as the primary foodstuff for embryos that grow from a few centimeters to more than a meter inside their mother's uterus. And in some marine snails, thousands of unfertilized eggs serve as bag lunches that stay fresh until needed for the lucky few embryos that hatch from fertilized eggs.

Polyembryony and New Roles for Marginal Offspring

Polyembryony is an intriguing route to parental optimism. The fission of a single egg results in a series of genetically identical clones, though genetically identical does not necessarily mean morphologically identical. In humans, monozygotic twins may appear physically different at birth due to prenatal differences in growth and development—Jacob and Esau, for example, who may have suffered from twin transfusion syndrome (see chapter 8). Differences become far more pronounced in many species, and such genetically identical offspring can be assigned specialized roles. Parasitoid wasps provide a striking example: a single egg of the encyrtid wasp *Copidosoma floridanum* injected into a host (usually the egg of a moth) divides into as many as three-thousand embryos, genetically identical but not phenotypically (morphologically, behaviorally) identical. A female egg divides into female larvae, a male egg into male larvae.

Most larvae take a month to complete development, but a minority develop precociously and quickly (development lasts about a week) into a warrior caste, armed with oversize jaws.

The precocious larvae of this and other parasitoid species serve dual roles. First, they defend their sibs from other parasitoid larvae, either predators or competitors, of other species or of other clones of the same species. Second, they regulate the sex ratio of mixed-sex broods. The wasps practice sib mating, and the number of males required to ensure complete fertilization of the sisters is small. Moreover, larval growth is a zero-sum game. More of the host tissue eaten by the males means less for female larvae. But the female parasitoid larvae deal effectively with surplus brothers using Amazon warriors that ruthlessly murders males.

Under the protection of warrior sibs, normal larvae are free to continue their assault on the host tissues. They consume the organs, muscles, glands, gut, nerve cord, and body fluid, and by the time they are finished the moth host is an empty shell packed with wasp larvae. The survivors pupate and gnaw their way to the outside, bursting through the host integument. Sound familiar? The warrior caste, though, does not complete the journey. They are sterile and never develop into adult wasps, forfeiting their lives to serve their genetically identical sibs.

Parasitoid Wasps

Parasitoid wasps inject eggs into living hosts—plants, aphids, caterpillars, spiders—where larvae grow and develop. Usually the parasitoid kills the host, eating it from the inside out before erupting in *Alien*-like fashion. Many parasitoids are solitary. Only a single larva ever survives even if more than one egg is laid. But even in solitary parasitoids the likelihood of successful parasitism rises with the presence of an extra egg. The battle between parasitoid and its host is life or death. The parasitoid must crack the host's immune defenses or surrender its life. The host must arrest the development of the invader or similarly forfeit its life. Hosts attempt to surround the parasitoid to cut off access to critical resources including air, and if successful win often by asphyxiating the larvae. But for solitary parasitoids there can be strength in numbers. Ichneumon wasps that parasitize weevils and moths enjoy greater success by laying multiple eggs. The ichneumon wasp *Hypeseter exiguae* parasitizes noctuid moths, and only one in twelve eggs laid singly ever survives. But when two eggs are laid, one larva survives in 97% of hosts. Only one of the two larvae will live on, but two eggs are better able to overcome host defenses than one. Here the transition from harmony to discord is rapid. Once the host defenses have been breached, the parasitoid larvae become mortal enemies.

Adaptive Suicide?

Suicide is not an obvious route to evolutionary immortality, but the notion that suicide might be adaptive has long held a dark fascination for behavioral ecologists. A quarter century ago the ornithologist Raymond O'Connor published a paper with a provocative title: "Brood reduction in birds: Selection for fratricide, infanticide, and suicide?" Adaptive suicide? O'Connor used the logic of inclusive fitness to show that—in theory—suicide could be favored under conditions of stringency if doing so liberated scarce resources that could be used by other progeny with better prospects for survival. The idea, though logical, always seemed implausible to me, at least for birds. Brood reduction in birds is almost always the direct outcome of fatal sibling rivalry, and the threshold for fratricide (aka siblicide) precedes that for suicide in O'Connor's model. If a nestling was ever tempted to leap from its nest for the good of its siblings, its sibs would probably already have pushed it over the side. But is suicide possible in other organisms?

Pea aphids spend their summers sipping plant juices and reproducing madly by parthenogenesis (asexual). Each aphid in these all-female populations gives birth to six to eight nymphs every day, building genetically identical clusters. These aphids are vulnerable to predators, such as ladybugs, and parasites such as braconid wasps that parasitize aphids by injecting eggs into still-living victims. Upon hatching, a wasp larva devours its unfortunate host from the inside out and, when it reaches maturity, bursts forth from the carcass to begin anew the search for more aphids to parasitize. It chooses those nearby, the identical copies of the now dead victim. Not only is the parasitized aphid doomed; it contains the seeds (OK, larvae) of destruction of its clone mates, the closest of possible relatives.

But the approach of an aphid predator such as a ladybug triggers a remarkable behavior—parasitized aphids leap to their death, and the death of their *Alien*-like resident, much more often than unparasitized aphids. The aphid's suicide serves to protect its clone mates from further wasp parasitism. Is this adaptive suicide? Perhaps, though the idea is controversial. The leap to death occurs only when a predator confronts the parasitized aphid. Why don't aphids simply leap to their death upon being parasitized? That would appear to be the logical solution, but there are two possible deterrents to such behavior.

First, already mature aphids continue to produce their own off-spring before succumbing to the parasite. Interestingly, older aphids with that prospect of reproduction are less prone to suicidal leaps than young aphids, not yet mature and not likely to become so be-fore the alien-invader completes its mission. As with Arthur Dent, knowing the moment of one's death can be liberating.

Second, natural selection can edit only existing behavioral varia-tions, and pea aphids have a very limited behavioral repertoire. Leaping from plants is normally something to be avoided because of the high risk of death, particularly in dry habitats, where most leapers quickly succumb to desiccation (and suicidal leapers are more common in such habitats). But when confronted by a preda-tor, the wingless and thus flightless aphids have only two options: run away (my particular favorite) or jump. Since jumping is nor-mally associated only with avoiding predators and is thus a self-pre-serving (albeit desperate) act, it is not surprising that antiparasite behavior (suicidal leaps) would be superimposed on this. This may be the best that natural selection can do, constrained by the existing behavioral variation.

Honeybees practice haplodiploid reproduction. Females hatch from fertilized eggs and are diploid (two complete sets of chromo-somes, one set from each parent). Males normally hatch from unfer-tilized eggs and are haploid (one set of chromosomes from mother). But occasionally, diploid males arise. Male honeybees normally do little work around the colony, but haploid males, are tolerated be-cause they are needed for reproduction. No such benefit exists for diploid males, because they are effectively sterile, and thus these in-dividuals represent a drain on the colony economy. Their solution? They "ring" a chemical dinner bell with an unusual message: come eat me! That is, diploid males produce a cannibalism-inducing sub-stance that causes workers to consume them. They appear to collab-orate in their own consumption!

Termites manufacture a soldier caste whose members are larger and more formidably armed than other workers. Soldiers of nasute ter-mites defend their brethren by playing Spiderman. They entangle their enemies with a sticky thread that they fire from a conical organ—a gun—that sits upon their head. But soldiers of the termite *Globitermes sulphureus* carry their valor to a suicidal extreme. They carry a reservoir of yellow liquid in their abdomen that they spray upon their victims and themselves, entangling both, with forceful contractions of the abdominal wall. On occasion the contractions of the defending soldier are so powerful that the defender explodes, perishing in a final heroic blast to protect its colony mates and relatives.

The Benefits of Teamwork

> *Two are better than one, because they have a good reward for their labor. For if they fall, the one will lift up his fellow, but woe to him that is alone when he falleth; for he hath not another to help him up. . . . And if one prevail against him, two shall withstand him; and a three-fold cord is not quickly broken.*
>
> —Ecclesiastes 4:9–10, 12

The advantages of social living are manifold—cooperative hunting and foraging; enhanced vigilance for or defense against predators; physiological or energetic benefits. Lions hunting in groups can kill larger prey; pelicans foraging in flocks can herd small fish into easy-to-catch balls. And even two can be better than one. Bald eagle parents will cooperate to become an effective hunting team. I was witness to one remarkable example when a pair of eagles from a nearby nest came across an unfortunate mother merganser (a fish-eating duck) and her brood. The eagles immediately chased off mother and then proceeded to systematically raid her brood. The eagles would fly over the brood from opposite directions and make carefully timed dives. The first eagle would swoop lazily on the ducklings, making no serious attempt to capture them. This was designed obviously to cause the ducklings to dive underwater. The second eagle, in tight coordination, then made its pass to coincide exactly with the exhausted ducklings popping up to the surface, where they were easily snatched. Within a quarter hour the eagles had plundered half the brood, neatly illustrating the benefits of teamwork.

Nature teems with such tales of animal teams, particularly among family members, among whom the barrier to cooperation is lowered by shared genetic interests. This pair of eagles worked in tandem to raise a family of eaglets at a nearby nest, and the payoff was immediately obvious: more efficient foraging translated into more babies, and more babies meant an enhanced evolutionary legacy. We see the benefits of teamwork clearly in social animals, and most clearly in social insects.

Social Insects: The Ultimate Team Players

Family harmony reaches its apex in the eusocial insects—ants, bees, and termites—that build and inhabit elaborate colonies. Their gigantic extended families are an ergonomic tour de force that even Henry Ford might have admired. Separate castes with distinctive morphology and behavior play well-defined roles within the colony superstructure. Oversize queens devote themselves to egg production, are doted on by all-female workers, and are protected by all-female soldiers. Indeed, excepting perhaps humans, the social insects exhibit the most elaborate and structured societies in the natural world.

Darwin was puzzled that workers and soldiers are effectively sterile, foregoing their own reproduction for that of their queen and mother. Although the story is not quite so tidy (a few workers lay eggs that survive), this model of efficiency rests on Hamilton's observation that one can promote one's evolutionary interests by promoting the reproduction of collateral relatives. Social insects do this with economies of scale and the efficiencies afforded by specialization. Among the ants, for example, there are three basic castes of females (males for the most part do little on behalf of the colony): workers, soldiers, and queens, and many variations exist within the first two. The workers and soldiers represent effectively sterile castes that sacrifice their own reproduction for the greater good. Queen honeybees, far larger than the workers, are egg-laying machines. Worker bees care for young, store food, and maintain the comb of the hive, among other household duties.

Intriguingly, workers, soldiers, and queens begin as genetic equals, but postfertilization events determine the role an individual will play in a colony: as one of the few to reproduce, or one of the many who do not and serve the needs of the lucky few. The many are the indentured workers, whose lives belong to their parents and siblings. They are the offspring served as food; the warriors who defend their brothers and sisters from parasites, pathogens, and predators; and the assassins who slay brothers on behalf of the sisterhood.

A key feature of this system in the ants, wasps, and bees is haplodiploid reproduction. Males hatch from unfertilized eggs and are haploid. Females hatch from fertilized eggs and are diploid. Sisters are in effect supersibs—they have the usual 50% probability of sharing one of their mother's chromosomes, but where there is a single father (sometimes there is more than one), the probability that they will share their father's chromosome is 100%. This genetic oddity results in a coefficient of relatedness between sisters of 0.75 instead

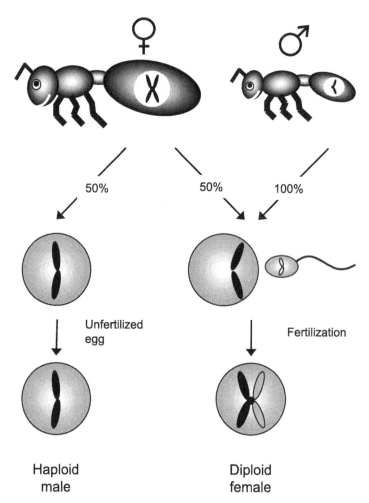

Figure 10.1. Haplodiploid reproduction
Social ants and bees practice haplodiploid reproduction, which results in unusual relational asymmetries among sisters, between brothers and sisters, and between parents and offspring. Males hatch from unfertilized eggs and are haploid. Females hatch from fertilized eggs and are diploid. Sisters have the usual 50% probability of sharing one of their mother's chromosomes, but where there is a single father (sometimes there is more than one), the probability that they will share their father's chromosome is 100%. This genetic oddity results in a coefficient of relatedness between sisters of 0.75 instead of the usual 0.5 between normal full sibs. The coefficient of relatedness between brother and sister is only 0.25.

of the usual 0.5 between normal full sibs (the coefficient of relatedness between brother and sister is only 0.25). This high degree of relatedness is not necessary for the evolution of sociality and sterile castes in insects, but it probably helps lubricate the engine of cooperation. The resultant division of labor and congruency of purpose is unparalleled within the animal world. A key disincentive to female workers producing their own children is the curious genetic asymmetry. A worker is more closely related to her sisters than she would be to her own progeny (as long as there is a single father—if the queen mates multiple times, this genetic quirk rapidly dissipates). Queens not only produce more sisters but also future queens that are also sisters to the workers.

The collective success of social-insect colonies is now so great that it stands as a barrier to individual initiative. In ants and bees, female workers give up personal reproduction to assist other workers and raise the progeny of their queen and mother. Workers are potentially capable of laying their own eggs and raising their own offspring, but within the colony the queen punishes such reproductive independence. A worker could conceivably leave the colony to strike out on her own, but such behavior simply does not pay, for a plethora of reasons. A major deterrent is that individuals cannot take advantage of the economies of scale and efficiencies of specialization afforded by the colony. At one time in the evolutionary past, prior to the evolution of elaborate sociality, individuals could reproduce independently and successfully, but in the modern world of the ant or bee collective, the little guy (or gal) can no longer compete. The two Steves, Wozniak and Jobs, at one time built state-of-the-art microcomputers in a garage. But with the rapid evolution of the personal computer industry, even those two wunderkinds could not compete with today's corporations arrayed with armies of programmers and factory production of hardware. There is no return to a simpler time, and as it is with the founders of Apple Computer, so it is with social insects.

The social insects represent the ultimate division between core offspring (queens, drones) and marginal offspring (workers, soldiers). Here the sole function of marginal progeny is facilitation, but it is facilitation taken to spectacular extremes. Ant collectives farm fungus, milk aphids, and even store honey within living bodies that swell to the size of a grape. Legions of workers clip leaves with robotic efficiency; others steal babies from other colonies to turn them into slaves. Still others carry specialized weaponry in the form of giant mandibles to defend colonies against intruders. All these tasks are performed by individuals who forgo their own reproduction to help the family produce more babies. Freed from the need to breed, these individuals can be morphologically specialized for their roles as support workers.

Trophic Offspring

Social insects often use their own eggs and larvae as living food stores. Indeed many animals—ranging from marine snails to sharks and owls—do the same. Eggs and larvae can be hors d'oeuvres as well as offspring. Some ants produce omelets of multiple fused oocytes designed for later consumption. Mother sand tiger sharks use unfertilized "trophic" eggs to feed a pup that grows from a few centimeters to more than a meter in length. And there are further examples that I shall explore in the next chapter on cannibalism.

The consumption of progeny ranges from rare, as in hawks and owls during periods of acute food stress, to routine, as in ants and bees. Eggs and larvae are the ultimate fast food for parents and siblings—nutritious, and they stay fresh until needed. This analogy was not lost on Richard Alexander when he coined the term "icebox hypothesis" to explain trophic offspring.

Sibling Synergies in Birds and Mammals

> *Again if two lie together, then they have heat, but how can one be warm alone?*
> —Ecclesiastes 4:11

With rare exceptions such as naked mole rats, the levels of cooperation within vertebrate families do not approach those observed in social insects. The advantages of having siblings in birds and mammals are more subtle. Birds and mammals are homeotherms, meaning that they regulate body temperatures physiologically within relatively narrow ranges. This allows for increased metabolic efficiency, as most of their biochemical pathways operate best within a narrow band of temperature. This key feature differentiates the birds and mammals from reptiles, amphibians, fish, and invertebrates. It is an energetically expensive lifestyle, and places practical limits on the numbers of offspring that can be raised. Except in precocial birds, which require little posthatching care, few birds and mammals raise more than a dozen offspring at a time. Homeotherms need to stay warm (but not too warm) to stay alive, and need food to fuel the process. And here is an obvious place to look for sibling synergies in birds and mammals.

Red-winged blackbirds are the archetypal songbirds. They build nests in wetlands and fields over most of North America and raise broods of one to four, occasionally five, chicks over the span of about three and a half weeks. Their development is termed altricial—the nestlings hatch naked and helpless, and require near constant parental attention (chiefly by mom) when young. Unlike precocial ducks and geese, whose chicks are self-feeding and mobile from hatching, altricial songbirds are high-maintenance progeny. Altricial blackbird nestlings betray their reptilian ancestry by beginning life as ectotherms (cold-blooded), relying on external heat for warmth and survival. Only gradually do they make the transition to endothermy or thermal independence, when they can regulate their body temperature within narrow bounds. During their cold-blooded phase, which extends roughly halfway through the nestling period, which lasts a little less than a fortnight, red-winged blackbird nestlings rely on their mother for warmth. She heats her offspring using a brood patch, a richly vascularized tissue on her breast. But mother blackbirds are effectively single parents (father is nearby but performs little direct care of the eggs or chicks), and a mother must leave the nest at regular intervals to search for food for both herself and her brood. During these absences her chicks cool; thus mother faces a balancing act between the search for food and the need to brood. Here siblings play a key role in heat conservation for the brood.

A brood huddled together reduces the exposed surface area of skin and rate of heat loss. Hatchling blackbirds are partially covered in down, but substantial feathering takes another week. And for nestling birds (and mammals) heat is every bit as important as food—more so, even, for a simple reason. Cold chicks cannot digest food. If the digestive tract and critical enzymes are not at operating temperature, it matters not how much food a chick ingests. Without heat, that food cannot be processed. Without digestion, growth and maintenance is impossible; death soon follows.

For red-winged blackbirds, three seems to be just the right number of chicks from a thermal perspective. There is little thermal advantage to more (and food costs increase). Below three, the chicks lose heat rapidly during the mother's absences, raising substantially the costs of heating and reducing the efficiency of food processing. But with a trio of ectothermic nestlings, the brood stays sufficiently warm, particularly when mother leaves the nest to search for food. An intriguing property of this process of development is that broods reach the state of endothermy before individuals. The transition from ectothermy to endothermy is gradual, and there are two advantages to being part of a brood: reduced heat loss from huddling, and the warmth produced by brood mates. As my ex-doctoral student Barb Glassey has observed, red-winged blackbirds seem to design their families not to maximize the number of surviving nestlings but instead to avoid broods too small, that is, below the critical brood size, particularly when the broods are young. Parents with clutches of three, four, and five eggs are all most likely to end up with three nestlings that survive to leave the nest. Like social insects, blackbird parents enjoy economies of scale. Small broods lose heat too rapidly, raising the energetic cost of thermoregulation. Large broods, though easy to keep warm, are difficult to keep fed. As with Goldilocks, an intermediate solution is optimal.

Conflict When Necessary, but Not Necessarily Conflict

The concept of parent-offspring conflict has received so much attention that one might be forgiven for assuming that enmity is the natural state of affairs. The underlying logic is impeccable. There are fundamental genetic asymmetries at play, and these can and do generate conflict. But it is far from inevitable. The behavioral ecologist Roger Evans elegantly demonstrated this simple point in a remarkable study of incubation in herring gulls.

Parent-offspring conflict or cooperation is notoriously difficult to study because the basic questions—what do parents or offspring want—are so hard to answer. Evans found an ingenious solution to this problem. He asked parent and offspring herring gulls directly. Incubation in birds is critical to successful development, as the eggs cannot stay warm by themselves. Below a lower threshold, embryo damage occurs, which neither parent nor offspring desire, and in many birds, as the eggs approach hatching, the embryos begin to signal temperature preference to their parents: chilled eggs peep, and parents incubate to rewarm them.

Parents and offspring could, however, disagree over how much warming is needed. Incubation requires energy expenditure, and one can imagine that parents might be less spendthrift than offspring desire. If true, a conflict arises: eggs would prefer a higher incubation thermostat than parents. The problem with much of parent-offspring conflict theory is that this is the default view. Conflict is assumed until proven otherwise. Roger Evans removed all doubt by asking parents and eggs a simple direct question: what do you prefer?

Evans allowed parent gulls to set their own preferred temperature over artificial eggs, and they chose 33.9°C, or a bit above the lower limit for normal development. The ingenious part of the study concerned the eggs. A water bath with hot and cold water inputs was set beneath the eggs, and a microphone was placed nearby with a link to the temperature control. Peeps from the eggs would warm the water bath, and thus allow the eggs to set their own incubation temperature. The temperature the eggs chose was also just above the lower limit for developmental problems, and in fact was slightly lower than that chosen by parents. Here parents and offspring agreed on the appropriate incubation temperature, and the signal to parents was simple and honest. This is perhaps the best empirical study of conflict theory for any bird or mammal: the verdict is an absence of conflict.

In fact, the observation that offspring chose a lower temperature hints at another key property of the communication system. Offspring have better information about whether they are too cold than parents. Without this good information, parents seem to err on the slightly high side. As the cost of erring low (dead offspring) is greater than the cost of being high (a slight increase in energy expended), the direction of the effect is not surprising. It also suggests not only that conflict is absent but that the system works more efficiently when the offspring are in control.

The mathematical theory of parent-offspring conflict is generally founded on the twin assumptions of increasing benefits of additional parental investment, albeit with diminishing returns, and increasing costs with additional investment—that is, a curtailed ability to invest in future offspring. Both assumptions are entirely plausible. But neither assumption can hold for the preference of incubation temperature in herring gulls. If the benefits are ever increasing, and costs continue to rise with further investment, then offspring will always favor a greater investment than parents. If, however, the costs remain flat as investment rises, both parents and offspring should favor an infinite investment that is biologically nonsensical. But if the payoff for additional investment rises to a maximum and falls off thereafter (a dome-shaped curve), the story grows more interesting.

Birds and mammals are homeotherms. Maintaining a stable body temperature allows for efficient metabolic functioning, as most metabolic pathways rely on enzymes that function best within narrow temperature ranges. Efficiency thus follows a "Goldilocks" function: too hot or too cold is bad; an intermediate temperature is just right.

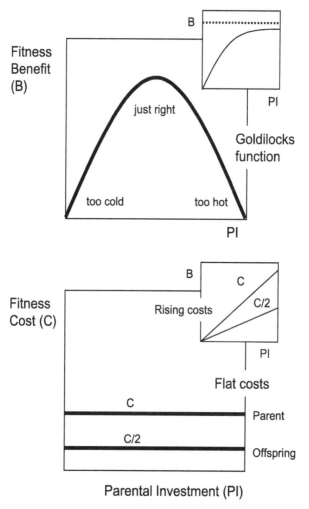

Figure 10.2. Parent-offspring conflict or cooperation?
Standard models of parent-offspring conflict assume that benefits (B) to off-spring rise with increasing parental investment (PI), albeit with diminishing returns, according to a Smith-Fretwell curve (*see inset in top panel on this page*). But such is not always the case. Consider incubation or brooding temperature. Temperature-based systems operate at peak efficiency within a narrow thermal range. There is, for example, a thermal optimum for the growth of nestling birds and litters of mammals. Temperatures too high or too low impair growth; intermediate temperatures are optimal. The *top panel* illustrates such a "Goldilocks" function.

Conventional models of parent-offspring conflict also assume that the cost (C) of parental investment (foregone future offspring) rises with the quantity of parental investment (*see inset in bottom panel on this page*). However, brooding or incubation is relatively cheap, so here I depict the costs as flat (*bottom panel*). As offspring are only related to future full siblings by a coefficient of relatedness of 1/2 and parents are equally related to all their progeny, offspring discount the cost of investment by 1/2 (C/2 for offspring vs. C for parents).

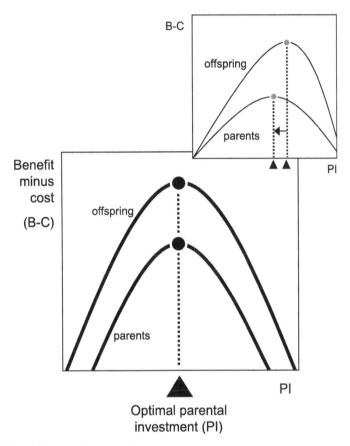

(*Continued from previous page*)

The net benefit of investment is the difference between the benefit and cost, which for parents if B – C and for offspring is B – C/2. Under conventional models of parent-offspring conflict, parents are expected to be more conservative in their investment in current offspring than offspring would desire—that is, the optimum investment from the offspring perspective is greater than that from the parent perspective (*inset in panel above*). But with a Goldilocks function and flat costs of investment, parents and offspring can agree on the optimum investment. The work of Roger Evans and his students on incubation in herring gulls found exactly such cooperation between parents and offspring.

The payoff for extra investment in incubation follows such a Goldilocks function. Eggs kept too hot perish. Eggs too cold will develop
abnormally or not at all. This is the first pillar of cooperation between parents and offspring in this system. But it is not enough to
generate cooperation: if the cost of incubation rises with increased
temperature, which seems plausible (parents must expend more energy), then parents will still prefer a lower temperature than offspring. The only way that parents and offspring can agree is if the
costs do not rise with increased parental investment.

Physiologists can make careful measurements of incubation, and
indeed, warmer temperatures require more calories. But is this truly
more costly, or is the extra energy expended trivial? My mailbox is
about 50 meters from my house, near the end of the street. If it were
at the other end of the block, the distance would be about 150 meters. Would it make any significant difference to me if I had to walk
the extra 100 meters each way to pick up mail? Probably not. The
cost in time and energy would be very small, and unless both were
in very short supply (which they are not), the extra investment
would result in no demonstrable cost. For a healthy adult bird, a
small rise in incubation temperature is probably like adding a few
meters to a daily walk: costs do rise, but only negligibly. If in fact
the cost increase for parent herring gulls for a higher incubation
temperature is trivial, then parents and offspring can agree on incubation temperature.

Evans's result is remarkable in two respects, one positive, the other
not. The positive aspect is that he asked parents and offspring to
set their own optima and obtained a tidy result. The result has far-
reaching implications for conflict theory and the study of parent-
offspring communication systems. The second striking aspect of this
work is that few others seem to have noticed. As of this writing (December 2003), the paper has been cited by other scientists in the
field a meager five times.

Cooperative Defense . . . against Parents?

Sibling birds and mammals are often rivals for food, and the outcome is frequently fatal. With the death of one offspring, more food
is liberated for the survivors, creating a powerful incentive for brood
reduction. Indeed, this is a potential source of conflict between parents and offspring. For parents, who are equally related to all their
progeny, the optimal brood size is generally going to be larger than
that for an individual offspring, who is more related to itself than

to its sibs (except where offspring are clones). Not surprisingly, when brood reduction occurs, it is usually the proximate outcome of fatal sibling rivalry. The most extreme form is siblicide. But parents are not entirely disinterested observers in this game. Should a brood be reduced to an unprofitably small size, parents may be tempted to abandon the current breeding attempt in favor of starting anew. This threat of parental abandonment may be a powerful incentive for siblings, and in particular older, stronger siblings, to treat younger sibs more generously.

Great egrets and great blue herons are large wading birds that raise broods of two to seven nestlings on diets of fish, frogs, and invertebrates. The nestlings are siblicidal, and their spearlike bills, ideally designed to capture prey, are also serviceable weapons for poking sibs, as my colleague Douglas Mock has reported in great detail. Siblicide serves to reduce competition for food in these birds; fewer competitors mean a larger per capita food share for the survivors. But here the selfish tendencies of sibs are checked by enlightened self-interest. Doubleton broods are far less prone to brood reduction, especially early in the breeding season. The reason appears to be that parents left with just a single chick, and an opportunity to begin breeding anew, will simply abandon the lone offspring in favor of raising a larger brood later in the season. As Mock and his coworkers have noted, an egret nestling that eliminates its lone remaining sibling from a doubleton brood is tempting fate; thus the presence of the extra sibling may serve to ensure continued parental care.

Facilitation in Humans?

Eros needed the reciprocal love of his brother Anterios before he could thrive. A prenatal parallel to this Greek myth may occur in humans. The incidence of dizygotic (two eggs) twinning rises in mothers in their late thirties and then falls thereafter with the approach of menopause. As mothers grow older, their uterus becomes less and less receptive to fertilized eggs, and the likelihood of pregnancy falls, a fact not lost on practitioners of in vitro fertilization. This procedure involves the administration of fertility drugs to stimulate mothers to produce abundant eggs, which are then harvested, fertilized in a petri dish, and returned to the mother. For mothers over forty, success rates are very low with a mother's own eggs, as the quality of the embryo shows a dramatic decline over and above the declining receptivity of the uterus. Because of this, multiple embryos (two to seven or more) are customarily transferred in the procedure, elevating the risk of multiple pregnancy (see chapter 7 for a further discussion). But there is a curious property about the statistics of these transfers. The pregnancy rates for transfers with multiple embryos soar over those for single embryos in women of all ages. Pregnancy rates for two-egg transfers were 2.1 to 2.4 times higher than for single-egg transfers, and rates for three-egg transfers were 3.6 to 4.0 times higher. Transferring more eggs most assuredly increases the chance of producing at least one good embryo (a progeny choice benefit of parental optimism), but this alone cannot account for the observed pattern. It could be that mothers who had only a single embryo transferred were less fertile than mothers with multiple embryo transfers. But perhaps the presence of a second embryo facilitates the process of implantation, and a plausible mechanism immediately suggests itself: the joint production of pregnancy hormones such as HCG (human chorionic gonadotropin). HCG is a prerequisite for early pregnancy maintenance and is produced before, during, and after implantation (implantation occurs about one week after fertilization). Menstruation resumes, and spontaneous abortion occurs, if an embryo does not secrete a sufficient level of HCG (see chapters 4 and 6 for more detailed discussion). Two embryos produce more HCG than one. Might two embryos be better able to avert spontaneous abortion than one? Perhaps. It would seem that this question is worth further investigation.

Finding Their Niche: Birth Order and Human Behavior

Under a strategy of parental optimism, parents create surplus progeny and then play favorites within the brood. Rank within the brood becomes a powerful determinant of the behavior of offspring, in part because they are assigned different roles by parents and in part because offspring of different rank confront differing opportunities. Human progeny are not exempt from the rules of parental optimism, and primogeniture is one such example. With the family property passed to the eldest son, older and younger progeny are socialized within families for their future roles. Primogeniture arises where land is critical to economic success and in short supply, enhancing the survival of a least one line of the family across generations, a measure that probably tracks evolutionary fitness. Parental favoritism thus encourages diversity within the family: if the eldest son is to receive the farm, younger siblings need to find other occupations. Parental favoritism has played a key role over the course of human history. Primogeniture in Viking society forced younger sons to seek their fortunes overseas, and between the eighth and eleventh centuries Vikings expanded their trade routes in the south, east, and west.

This theme is explored at length in Frank Sulloway's *Born to Rebel*. Sulloway's thesis is that the family structures interactions between parents and offspring, and among the offspring themselves. Siblings compete for parental attention and, more important, parental investment, and are sensitive to parental favoritism. Firstborns align themselves closely with their parents' needs and interests, and force later-born siblings to work harder and diversify to succeed. These role-dependent opportunities and constraints create niches within the family that can even overshadow variation in environmental conditions across families. The behavioral/personality differences observed within families between older and younger siblings are almost as great as those across unrelated individuals, with firstborns being conscientious, goal oriented, status seeking, and fierce defenders of the conservative status quo. Later-borns are more likely to rebel against their parents' values, are more open to new experiences than firstborns, and are more likely to negotiate than to argue. This follows from the principle of diversification in Sulloway's view, and he supports his thesis with mountainous empirical evidence, compelling overall, though perhaps stretched too thinly in places. Firstborns and later-borns occupy different niches within the family, and this has far-reaching behavioral implications: scientific, political, and cultural revolutions are more likely to be born in the minds of, and supported by, later-borns. Only when the heterodox becomes orthodox are firstborns likely to jump on board. Sulloway traces these broad patterns back to their roots in the dynamics of families and the search for greater parental investment.

Sulloway codifies the principle of diversification, borrowed from Darwin, in four rules: (1) If you are born later, diversify. (2) If your parents have limited resources, diversify. (3) Diversify in proportion to the number of your siblings. (4) Under certain circumstances, disregard the previous three rules and specialize. The first three rules form a logical trio, whereas the fourth rule, a critic might argue, is an ad hoc explanation for exceptions to the first three and weakens the overall structure. This flaw notwithstanding, one can find broad support for Sulloway's themes in nonhuman nature.

Differences within families can be every bit as important as those across families. One need look no further than a brood of blackbirds. The early-borns—the core nestlings—enjoy unfettered access to food, whereas the later-hatched and marginal nestlings face an uphill battle with their older sibs. When food is short, marginal sibs are first to suffer the effects of the shortfall. As in humans, the differences within families exceed the differences across families. The survival of core nestlings is more or less assured: even in the worst environmental conditions, the average rate of survival of core nestlings from hatching to fledging never dips below 85% in red-winged blackbirds (and normally sits well above 90%). This result holds true across families of different size and years of different quality. For later-hatched marginal progeny, average survival rates range between 0 and 80%, with the lowest values experienced by last-hatched nestlings in large families in the worst years. Only in small broods in the best years does the average survival rate of marginal nestlings approach that of the favored core progeny. Marginal chicks must work harder for a lesser reward in blackbird families. Such behavior finds an exact parallel in human families: younger sibs must overcome their elder brother's or sister's monopoly on parental attention. Sulloway suggests that younger children deliberately avoid overlapping older sibs' interests and occupations in order to maximize parental investment and escape childhood with the best possible prospects for future success.

Sulloway's principle of diversification also holds in many animal systems. In siblicidal blue-footed boobies, younger sibs learn to become subordinate and thus reduce the costs of conflict over the period of parental care. And as we have seen above, some nonhuman organisms take this process even further with the Arthur Dent effect. With the door of future reproduction closed, marginal offspring in social insects diversify into supporting roles for parents and older siblings—soldiers in social insects, or warrior offspring in parasitoid wasps, or trophic offspring in bees, frogs, and sharks. These are offspring that serve their families best by dying.

Chapter 11

Cannibalism and Infanticide

*I grant this food will be somewhat dear, and therefore very
proper for landlords, who as they have already devoured most
of the parents, seem to have the best title to the children.*
                                                    —Jonathan Swift

The Pathways to Cannibalism

In 1729 Jonathan Swift made "A Modest Proposal for Preventing
the Children of Poor People in Ireland from Being a Burden to their
Parents or Country, and for Making Them Beneficial to the Public."
Swift's solution? Fatten young babes on mother's milk and then, at
a year, slaughter and eat them. At ten shillings per child, a fee paid
by persons of quality and fortune, mothers could earn a handsome
profit, as Swift estimated the rearing costs for each child to be no
more than two shillings. What he suggested was of course dark satire, a bitter commentary on life in eighteenth-century Ireland,
where so many children were born to short, bleak lives of desperate
poverty while feudal landlords lived in comparative luxury.

Swift's satire works so well because we expect warm and loving
relationships between parents and offspring: we are unprepared for
the polar opposite. Infanticide is not an obvious route to enhancing
fitness yet is widespread among both plants and animals. And com-
pounding the sin by cannibalism presents an even greater affront
to human sensibilities. Swift accuses landowners of the figurative
cannibalism of their tenants, but the consumption of infants occurs
routinely throughout the animal kingdom. Indeed there is nothing
particularly special about eating someone else's children, as one parent's offspring is another's hors d'oeuvres. Filial cannibalism—the
consumption of one's own offspring—is another matter. And parental optimism provides another pathway to cannibalism: parents create the problem of having too many children around the dinner
table and then solve it by dining on the kids.

Honey, I Ate the Kids

In spring and early summer, male largemouth bass clear a nest about
a meter across on lake bottoms in shallow water. They fan the incu-

bating eggs and guard the nest contents against all intruders. Dad's care is essential, for if you remove the father, the brood is likely to fail. When the eggs hatch, the swarm of fry remains under dad's watchful eye. Eventually the family breaks up, and the baby bass join the other little fish among the submerged weeds of the lakeshore. Later in the summer dad may again encounter his own progeny, but now his interests have changed from guardian to gastronomic. They are now just little fish, and little fish are good to eat.

The cardinalfish *Apogon doederleini*, is a resident of the western Pacific along rocky shores and reefs where it breeds. But unlike most fish, which either build nests or cast their eggs adrift after spawning, cardinalfish are mouthbrooders. Immediately after spawning, the male picks up the egg mass in his mouth, where it stays until the eggs have hatched and the offspring are away. In doing so, fathers protect their progeny against predators and pathogens, and ensure adequate oxygen for the developing eggs. But this paternal care comes at a cost: fathers must fast while mouthbrooding. For his offspring, the father is both devoted kin and potential cannibal. How many eggs a male can care for depends on the size of the father's mouth cavity, and mothers, as is the custom, are sometimes optimistic about their clutch size. Too many eggs means not enough oxygen for all, a problem that dad solves simply by snacking on the surplus. This is partial brood cannibalism. But the male may also consume the entire brood if his condition should decline during the mouthbrooding fast, particularly if new mates are abundant. If a male can replace the lost clutch rapidly, he loses little time and gains a valuable meal by cannibalizing his brood, and his progeny become provisions. One wonders whether the little cardinalfish find themselves in need of later counseling to cope with their father's split personality.

A gull colony is a raucous and busy place. Nests are built on the ground and densely packed. Territorial boundaries are sharply defined, and often extend no more than the bill's reach of an incubating parent. After hatching, a chick enjoys its parents' protection while inside the territory. But stray beyond this invisible line, and life takes a perilous turn. A wandering chick is sure to be battered by other adults with no interest in foster children and may become a snack for another gull that sees cannibalism as an opportunity for an easy meal. Yet some chicks do run away from home. They are almost always the last to hatch from a trio of eggs, and underfed. Stay at home, and starvation is near certain. These chicks make the best of a bad job and head off. Their goal? To crash another brood of hopefully younger chicks, where their role would switch from runt to big brother or sister. But the odds against our wandering chick are steep. Other parents are wise to this game and do their best to repel the intruders. A chick on the outside is an interloper, a potential drain on Darwinian fitness, as raising orphans is expensive and yields no evolutionary payoff. Occasionally though, a parent is—quite literally—caught napping, and our hopeful chick slips through the territory perimeter. Its prospects improve markedly as it is fed alongside other chicks in its new family.

This simple decision rule of parents—be kind to chicks inside but not outside the territory—sometimes misfires. One would-be cannibal gull raided chicks from other nests and returned to its own nest presumably with the intent of eating them. But once chicks cross the invisible boundary of the home territory, they become children instead of cuisine. As our cannibal gull persisted in its raids, the consequence was an ever growing family.

Consuming one's own children would seem inherently wasteful. Even Swift would have rejected such a notion, as it would quite literally eat into one's profits. But just such ghoulish behavior occurs under the cloak of darkness in an underground burial crypt. The perpetrators begin by locating a corpse and then excavate a burial chamber. The body is carefully prepared, with all body hair removed, and then deposited in its final resting place, where the cadaver is then guarded by one or often two sentinels. This body is not the object of cannibalism but a lure that brings the victims to the cannibals. Even more ghastly is that the intended victims are the sentinel's own children, babies that have just dug their way toward the body on which they hope to feast. Within a day of their arrival these children will be seized and devoured by the same individuals that gave them life, their mother and father. Such a fiendish plot could comfortably reside within the pages of a Stephen King novel, but it is quite real, part of the normal cycle of reproduction in burying beetles of the genus *Nicrophorus*.

These beetles entomb the carcasses of small vertebrates such as mice in a burial crypt and then lay their eggs in the surrounding soil. Upon hatching, the larvae make their way to the body, where they feed and grow. But burying beetles, like so many plants and animals, are optimistic parents. The number of larvae exceeds the available food supply, particularly when the carcass is small, necessitating parental intervention. And that intervention is cannibalism. Parents cull their brood to ensure that a small number of robust progeny survive instead of a larger number of malnourished offspring.

The surplus offspring are a hedge against the uncertain survival of progeny. Not all eggs hatch, nor do all hatched offspring make it to the corpse, some being intercepted by predatory insects or falling victim to fungal infection. The excess progeny ensure that parents will not be left with a brood too small. Cannibalism is less frequent on larger carcasses, where there is ample food for all, rendering a cull of offspring unnecessary. How parents decide how to balance offspring number against available food—the corpse—was revealed in a series of clever experiments by the behavioral ecologist Curtis Creighton, who manipulated carcass mass by adding lead weights. Beetles who "thought" they had a larger carcass laid more eggs.

Infanticide also arises in these beetles for a very different reason. Not only do they consume their own progeny; they will eat the progeny of other beetles as an outcome of sexual selection. The entombed corpse is a valuable commodity for raising burying beetle babies, and the resident parent or parents must defend it against other burying beetles who might try to usurp the carcass for their own progeny. If an interloper succeeds, it cannibalizes any larvae on the corpse and begins reproduction anew. Such cannibalism serves to protect the food source—the corpse—for its own progeny. It does not arise from a strategy of parental optimism but rather is the outcome of sexual competition between beetles. Infanticide either with or without cannibalism often occurs not for reasons of gastronomy but because doing so promotes an individual's reproductive agenda, as with burying beetles that seize a larvae-laden mouse carcass from other beetles.

Offspring Designed to be Eaten

Swift suggested that human children could be raised for human consumption at a profit. One wonders whether Swift was aware that many organisms within the animal kingdom—creatures as diverse as sharks, snails, and bees—do exactly that. Some ants, for example, produce fused ova designed for consumption. And after a sand tiger shark pup has murdered its living siblings in utero, it feasts on unfertilized eggs. Red whelks, a marine snail, create egg capsules with 5,000 eggs, but only one or two are fertilized. The other 4,998 or 4,999 are lunch for the lucky larvae.

Richard Alexander coined the term "icebox hypothesis" to suggest an adaptive function for such trophic offspring. These progeny represent a biological storage system, a food reserve that can be drawn upon when necessary. In some birds and mammals, the consumption of progeny is associated with acute food stress, as in certain hawks and owls. Cannibalizing the victim serves the dual function of providing a timely meal and, perhaps more important, ensuring that there is one less mouth to feed. Somewhat surprisingly, cannibalism is relatively rare in siblicidal birds, even though the victim could easily be consumed by either a parent or a sibling. This suggests rather strongly that the second function, reducing current and future demand, is generally the more important of the two functions.

In other animals, dining on viable progeny and nonviable trophic eggs is a more or less routine component of family planning. The trait has arisen many times, in most cases from simple oophagy (the consumption by parents or siblings of dead or dying eggs). But from such humble beginnings, elaborate systems of using offspring for food have arisen. Parent ants and bees and snails and sharks and frogs deliberately create surplus offspring, or at least unfertilized but nutritious eggs, for the explicit purpose of enhancing the development of other family members by serving as food. These progeny forfeit their lives so that their relatives will thrive.

But not all trophic offspring need lose their life to assist close relatives. Ants of the rare genus *Adetomyrma* practice a bizarre form of feeding: queens chew holes in their larvae and then consume the oozing fluid (hemolymph). The amount of scarring on larvae suggests that this nondestructive cannibalism is routine, earning these Madagascar insects the nickname "Dracula ant."

The Pathways to Infanticide

Cannibalism of one's own children is very different from cannibalism of someone else's. The latter requires no special explanation, but filial cannibalism is, at first glance, paradoxical. Recognizing that this is part of a strategy of parental optimism helps to resolve the paradox. Cannibalism is just a specific form of infanticide, and a similar dichotomy exists between filial infanticide and infanticide of nonkin. Indeed, these are very different phenomena that involve only the superficial similarity of killing infants. One kills someone else's infants for very different reasons than one kills one's own biological progeny.

Sexually Selected Infanticide

Lions in Serengeti National Park live in prides consisting of three to twelve adult females, one to six adult males, and variable numbers of cubs. The pride defends a territory, and normally all females within the pride are related. They were born there and typically stay with the pride to breed. Female lions reproduce at ages ranging from four to eighteen years. Males, though, leave their natal pride at age three and, after a period of nomadic existence, will normally band with other males, often unrelated, to take sexual ownership of a pride away from the resident males. A male's tenure with a pride and hence its opportunity to reproduce is temporary, lasting on average two to three years. Ultimately the resident males will be displaced by a new gang of younger, stronger males. Thus, a male's reproductive life span is typically short relative to that of the females, and males evicted from one pride may never breed again. This disparity between male and female reproduction helps to explain the diabolical behavior of male lions. Upon taking over a pride, the new male(s) ruthlessly kill the cubs they find with a quick, lethal, neck-breaking bite. The obvious question is, why? Male lions are not alone in this brutal behavior—something very similar occurs with the small mammalian coresidents of many human homes.

Killer Rodents

House mice live fast and die young. Their average lifespan is just one hundred days, but they make the most of their short life—a female mouse under favorable conditions can produce a litter of four to nine pups every month. Males participate in caring for the brood and are indeed doting parents; except for nursing they perform all the same duties as mom—cleaning and grooming the pups, building nests, warming the litter. But male mice are also ruthless, infanticidal killers—it just depends whose progeny we are talking about. His paternal affection extends only so far as his own offspring, not those of other sires. Should a male encounter pups that are not his own, he is likely to murder them with a swift, lethal bite to the head or body. The mother, now deprived of her offspring, accepts the father-to-be's sexual advances sooner rather than later, and since time is short for mice, the male enjoys his evolutionary reward.

Is such infanticide in the mother's interest? Clearly not. If she could prevent infanticide, she no doubt would. But she is a victim of circumstance and helpless to resist a determined killer male. Mother mouse loses this battle of the sexes. Female mice resort to spontaneous abortion of unborn pups if they even get a whiff of a strange male during pregnancy, a bizarre phenomenon known as the "Bruce effect," so named for Hilda Bruce, who first described this pregnancy block. This behavior is evidently a counterstrategy to male infanticide: a pregnant mouse makes the best of a bad situation by cutting her losses. The smell of a strange male is a reliable signal that the resident male is gone, and it means that her pups-to-be are doomed to become victims of infanticide at the hands (or paws) of the newcomer.

Langur monkeys and lions live in female-based bands or prides. Males take sexual ownership of these groups by defeating and evicting the resident males, a task that is complete when the infants are killed. Again females become victims of differences of evolutionary interest between mothers and fathers. But female lions and langurs are not completely helpless in this evolutionary game. Females, who are often close relatives, may band together with some success to protect their infants from infanticidal males. Females pregnant at the moment of a male takeover will come into a false estrus and copulate with the newcomers in an attempt to confuse paternity of the offspring-to-be and thereby deter the future stepfathers from killing the future babies.

Why are male lions, langurs, and mice infanticidal? The simple answer is that time is short. The death of the existing infants brings females into estrus sooner, accelerating the mating opportunities for the incoming males, who will enjoy sexual access to the females for only a limited duration. The average tenure for male lions before they are evicted by a new and stronger band of males is two to three years, just enough time to complete a reproductive cycle and rear cubs to independence. If they wait for a female to complete her parental duties with a previous male's offspring—to whom they are unrelated—they may father no offspring of their own. As they have no genetic stake in the success of the previous male's progeny, infanticide is the path of self-interest they follow. Again females are not disinterested observers in this evolutionary game and do their best to counter the male's pitiless behavior. Resisting the infanticidal attacks of larger, stronger males is one obvious counterstrategy, and female lions with older, mobile cubs employ another. Such a female steals away from the pride until her offspring reaches the age of independence, and she returns only when the cub can avoid infanticidal attacks. Coming into a false estrus is a third female counterstrategy to male infanticide; it seems to work because male lions are not good at telling time.

But not so the humble male house mouse. It possesses an ironclad defense against such female deception: a unique neural timing mechanism tied to copulation. Mating with a female mouse starts a clock ticking, and the alarm goes off about three weeks later. This duration, not surprisingly, corresponds to the gestation period of mice. Thus a male mouse will remain infanticidal for two and one-half weeks following copulation, and any pups that he encounters over this period are sure not to be his own. But as the three-week mark approaches, the male switches from infanticidal beast to devoted parent. This astonishing mechanism ensures that a male will kill only the progeny of other males but not his own genetic offspring. Female counterstrategies to cuckold a male into caring for unrelated progeny are thus doomed to fail. There is now no incentive for a pregnant female mouse to attempt to deceive a new male by coming into a false estrus. Instead, the Bruce effect. Female mice spontaneously abort pups that are doomed to become victims of infanticide, cutting their losses short by abandoning their unborn progeny and starting anew. Male mice, lions, and langurs make very bad stepfathers.

Infanticide in Families

Infanticide that occurs within families is not a single phenomenon. Infanticide of one's biological children arises for very different reasons than infanticide of stepchildren, and nowhere is this more clearly seen than within human families. Infanticide has always been and sadly still is a routine component of human behavior, ranging from simple abandonment or neglect of infants to the drowning or burial of newborns. And modern medical science has added to the array of methods used to dispose of unwanted but as yet unborn offspring with induced abortion.

Infanticide arises in humans for the same reasons that it arises in organisms as diverse as apple trees, ants, and aardvarks. There are three main pathways. The first is simple pathology. Until the rise of sociobiology in the 1970s this was considered as the chief if not the sole motive for infanticide: babykillers must be sick, and infanticide as an adaptation was, as Darwin suggested, a non sequitur. The second path to infanticide is as a component of a strategy of parental optimism. Some progeny are created in excess of the parents' immediate needs as a hedge against defective progeny, or food shortfalls. The residents of ancient Sparta quite literally threw defective progeny away. Pre- and postnatal infanticide is more common for twins than for singleton infants, allowing mothers to avoid the sometimes unaffordable burden of a family too large. Parental optimism explains filial infanticide but does not explain the infanticide of male mice, lions, and langurs. That is explained by sexual jealously, the third major path to infanticide.

The Unwilling Parent?

The evolutionary psychologists Martin Daly and Margo Wilson are perhaps the foremost experts on the evolution of homicide in humans, and much of their work has focused on infanticide and child abuse. They use the term "Cinderella effect" to describe the epidemiological relationship they observe in stepfamilies. Infanticide and child abuse are far more common in families with stepparents/children than in families with the genetic parents intact. Daly and Wilson argue that kin selection provides the explanation.

Stepmothers are portrayed as evil in folklore and nursery tales. This stereotype is grossly unfair, as most stepparents are *not* wicked and cruel, but kind and loving. But there is a grain of truth that underlies the stereotype. As a statistical observation, children fare better with their genetic parents than with stepparents. Daly and Wilson suggest that this is because of parental discrimination based on the rules of kin selection.

One of their first studies revealed that a stepchild in the United States in the mid-1970s was roughly seven times more likely to be a victim of child abuse than a child living with two genetic parents, and one hundred times more likely to be a victim of fatal child abuse. This relationship was not explained by differences in wealth—low-income families were no more likely to have stepparents than high-income families. Subsequent work on Canadian families showed the same pattern. And when child abuse in families with stepparents occured, it was normally directed toward the stepchildren. The patterns that Daly and Wilson identified have been found in the United Kingdom, Korea, Nigeria, Hong Kong, Australia, Paraguay, Finland, Columbia, Malaysia, and Trinidad, among first-world peoples and traditional hunter-gatherers.

If Daly and Wilson are correct, then we should see not only an elevated risk of infanticide in families with stepparents but also evidence of the sublethal effects, behavioral discrimination against stepchildren. And here the evidence falls neatly in place. Stepchildren disproportionately fall victim to child abuse, including sexual assault. Among Canadian families only 1 in 3,000 preschoolers living with two genetic parents was a reported victim of child abuse. That proportion is forty times higher for children with stepparents, about 1 in 75.

Stepparents are more likely to be perpetrators of child abuse than genetic parents, but it is perhaps unfortunate that the data most easily gathered to test sociobiological explanations—statistics on criminal behavior—focus on the dark side of parent-offspring relations. Child abuse, even in families with stepparents, is rare. Though the rate of child abuse was forty-fold higher, than for children living with genetic parents, still only 1 in 75 preschoolers with stepparents was a reported victim of child abuse.

Daly and Wilson also find strong evidence that filicide is much more likely to be associated with mental illness—murder-suicide is almost exclusively found in genetic, not stepparents, and diagnosed psychiatric conditions are prevalent among those who kill their genetic children but not among those who kill their stepchildren. These facts certainly suggest different pathways of causation for infanticide by step- and genetic parents.

The work of Daly and Wilson, though widely respected, is not without its harsh critics. Some of this is simply hysterical opposition to the notion that there is a genetic basis to human behavior. Humans clearly are not automatons, hardwired for inflexible responses to particular conditions, but there are deep-seated and inherited dispositions that are conditioned by experience. We need to understand rather than ignore or deny these. One of the unhappy truths is that we are quite capable of infanticide and child abuse, and some of the time this can be traced to deep evolutionary roots.

How does sexually selected infanticide in langurs, lions, and mice relate to the phenomenon of stepparents in humans? There is a clear parallel in that the inclusive fitness component or lack thereof is certainly common to all. But there are key differences too. Infanticide is a regular feature of reproduction in mice, lions, and langurs but rare in humans. Whether it was always rare in humans is unclear, but there is reason to suspect so. Mice, langurs, and lions form comparatively temporary sexual liaisons, and longer-term associations in humans may be a powerful disincentive for infanticide. Is a human mother more likely to mate with an individual who has killed her children? Probably not. Tolerance of a previous father's children may be the price of a long-term reproductive relationship. But that does not mean that a stepparent will treat stepkids and his or her own genetic children identically.

Given the serious problems it generates, infanticide by stepparents is very likely maladaptive, particularly in the context of modern families. Rather, infanticide and child abuse more likely arise as byproducts of evolved psychologies that generate parental favoritism based on bloodlines. That psychological inheritance coupled to a stressful environment, may trigger infanticide, best described as a symptom of what can happen when things go seriously wrong. Critics of Daly and Wilson's thesis note that most stepparents are not infanticidal or abusive, and argue that this simple fact trumps any sociobiological explanation. Here the antisociobiology crowd brings the bogeyman of genetic determinism into service. Human males, however, are not male mice. Human behavior arises from the complex dynamic of a sophisticated genetic program and an equally complex and unpredictable environment. Daly and Wilson argue that human parents play favorites according to the rules of kin selection, via deep-rooted evolved psychologies. If correct, humans face a built-in bias against children that are not their own. The epidemiological data support this view strongly, but we need to know much more. Why do some parents do a better job of overcoming any innate bias than others? Is it simply circumstance? If one lives amid comparative wealth, it might be easier to avoid uncharitable urges than if one lives in poverty, and poverty is clearly an important determinant of infanticidal behavior. The suggestion that evolved psychologies are important is not an end point for discussion but just a beginning. What are the triggers for such discriminatory behavior, and can we regulate or control these?

Infanticide in human stepfamilies is an extreme manifestation of stepparents playing favorites. It is not adaptive, but the underlying program that favors parental discrimination is, or at least was, adaptive. Modern environments pose new challenges to old evolutionary programming. Software upgrades for all organisms take time as environments change, because natural selection cannot look ahead: it responds only to environments encountered, present and past. The behavioral program that favors parental discrimination following bloodlines has been debugged over millennia, but there is always a lag between the current environment and the existing genetic program. With slowly changing environments, this mismatch will be barely noticeable, but it would be fair to suggest that the human milieu is rapidly changing; some human programming is bound to be obsolete. If there indeed exists a genetic predisposition for discrimination against stepchildren, we need to identify the triggers for the ill effects that are sometimes seen. In most cases these are avoided, and stepkids enjoy happy, healthy relationships with their stepparents. But sometimes not, and we need to know why.

That said, those who argue that natural selection has not had sufficient time to modify behavior in modern environments may be surprised: much of evolutionary biology over the last three decades has shown how rapidly significant evolutionary change can occur. Radical change in the life histories of fish can occur in just over a decade; even new species can arise almost instantly. And humans are not exempt from rapid change. Human birth weight has long been a textbook example of stabilizing selection, with natural selection against the extremes favoring intermediate sizes. Babies born both too large (they get stuck in the birth canal) and too small (complications associated with prematurity) do not survive as well as babies born at an intermediate size. Or did not. The last half of the twentieth century saw the widespread use of the caesarian section in many Western countries as a solution to babies too large. The result? The penalty for babies too large has been relaxed. With this check on birth weight removed, mothers may soon be giving birth to babies with heads the size of basketballs.

Chapter 12

Brave New Worlds

*Our early semi-human progenitors would not have practiced
infanticide, for the instincts of the lower animals are never
so perverted as to lead them regularly to destroy their own
offspring.*
　　　　　　　　—Charles Darwin, *The Descent of Man, and
Selection in Relation to Sex*

*The history of our knowledge about primate infanticide is in
many ways a parable for the biases and fallibility that plague
observational sciences: we discount the unimaginable and fail
to see what we do not expect.*
　　　　　　　　—Sarah Blaffer Hrdy, *The Woman That Never Evolved*

DARWIN WAS MISTAKEN about the existence of infanticide. He and almost all others who were to follow for the next century did not see
because they did not want to see. Like it or not, infanticide is part
of the human behavioral repertoire. This comes as no surprise to
evolutionary biologists. It is a trait we share with virtually all other
species that extend the umbrella of postzygotic parental care to their
offspring. The same principles that explain the rude habits of sharks
that devour siblings, pandas and bears that abandon their progeny,
and murderous rivalries of pelicans and eagles underpin parallel behavior in humans. Filial infanticide has occurred throughout the
course of human history and still occurs today—most visibly in rural
societies, but no population can truly claim to be free of this human
vice. Some just do a better job of concealing it than others. We might
achieve a higher state of denial by passing stringent laws to legislate
it from existence and then turn a blind eye to its occult occurrence,
disguised as accidental deaths or hidden under the rubric of sudden
infant death syndrome.

Human infanticide differs in the particulars, of course, from infanticide among what Darwin referred to as the "lower animals" (and
plants too). But its occurrence—the proximate outcome of either
sibling rivalry or parent-offspring conflict—can be traced to the
same trio of reasons that underpin filial infanticide across the full

breadth and width of nature. Resources are unpredictably short and/ or individual progeny are not worthy of continued investment, according to the economic calculus of genes. And foster children are sometimes maltreated—even lethally—by stepparents for the same reasons of sexual jealousy that explain infanticide in lions and langurs. As with animals, so with humans: families often bring out the worst in us.

This unpleasant state of affairs was long resisted by the scientific community (and still is) because it did not fit with our biased but false expectations of how families really work and, for some, because of blind adherence to ideologies that could not neatly accommodate the ugly truth about what lurks within human nature. Until the mid-1970s, infanticide in birds and mammals was considered, by default, as pathological behavior. What occurred in lower animals was not quite so worrisome. That insects ate their babies was not surprising, since they eat anything. But the infanticide-as-pathology view has been displaced by an evolutionary view. Infanticide can be and often is part of an adaptive reproductive strategy. That is not to say that infanticide per se is adaptive; often it is not. Rather it is a by-product of an adaptive strategy that leaves infanticide as the best of an unpredictably bad situation. And some cases of infanticide are indeed pathological. Andrea Yates, the Texas woman who drowned her five children during a psychotic episode, was and is a deeply disturbed woman. But these cases attract so much prurient interest exactly because they are so rare. Most cases of infanticide are not pathological, and we get nowhere in understanding and avoiding this grisly behavior if we persist in a state of collective denial. The default assumption of infanticide as illness is wrong.

To understand the behavior, we need to identify the conditions under which it is likely to occur and the evolutionary antecedents. Infanticide is more likely to occur after twin births in cultures across the world for this very reason. For many mothers, particularly in preindustrial societies, twins are an unaffordable burden. This higher likelihood of infanticide arises not as a component of a resource-tracking strategy but rather to correct a developmental accident (monozygotic twinning) or as a maladaptive by-product of a strategy of progeny choice. And progeny choice occurs in two different contexts: sex selection and birth defects. Throughout the course of human history birth defects have been a trigger for infanticide. The process was even formalized in ancient Sparta with a set of rules specifying which infants were to be reared and a committee of elders to enforce them, with the unfortunate victims to be cast away.

Sex-biased infanticide, directed more often toward daughters, also has an equally long history. Such behavior is sometimes overt (drowning, smothering, burial, or abandonment) but is more often covert, operating via neglect (e.g., failing to breast-feed or early weaning, insufficient medical care) resulting in elevated infant mortality. And where cultural rules make daughters more expensive to rear than sons, female babies are especially vulnerable to infanticide. The phenomenon has been examined most closely in South Asian cultures with well-established rules for the practice of this behavior.

For obvious reasons human infanticide is notoriously difficult to study. It is rarely observed directly, but demographers can detect it statistically with equal certainty. When there are only 75 baby girls for every 100 boys, something sinister is afoot. Where are the missing daughters? The human sex ratio is slightly male biased at birth, with 106 or 107 males to every 100 females being the norm. But in societies with a tradition of female infanticide, the surplus of males over females rises sharply. Guangdong and Hainan provinces in rural China, for example, showed ratios of 130 and 135 males to 100 females in the 2000 census, and similar numbers of males to females were reported from rural Indian states in the mid-1990s. The twentieth century introduced modern technology to the process of sex discrimination. Prenatal screening by amniocentesis and ultrasound now allows for prenatal screening for sex and birth defects via detection and abortion. Humans embrace the basic principles of parental optimism: they produce surplus progeny and then cull their progeny for the purposes of insurance, progeny choice, and to avoid broods too large. And now we have introduced modern technology to expand this practice.

Artificial Parental Optimism and Infanticide

In 1994 an article titled "Successful Outcome of Multifetal Reduction in a Pregnancy with 12 Live Fetuses" was published in the academic journal *Human Reproduction*. A twenty-eight-year-old woman who could not ovulate naturally was treated with fertility drugs that hyperstimulated her ovaries, producing a veritable blizzard of eggs. She conceived successfully, and at the seventh week of pregnancy twelve live fetuses were detected by ultrasound. Two weeks later twelve fetuses—eleven alive, one dead—were still present, and the following week three were killed (embryo reduction) by injection of potassium chloride to reduce the brood size to eight. Another week later seven live fetuses were still present, and the woman underwent a second embryo-reduction procedure, trimming her brood size from seven to four. In the twelfth week of pregnancy, she underwent a third embryo reduction; one more fetus was "sacrificed," leaving the brood size at three. No further reductions were performed, but during the twenty-fourth week of gestation one of the remaining triplets died. At thirty-two weeks the woman gave birth "successfully" to one live and two dead fetuses. More is not necessarily better and is often much, much worse.

For women suffering from infertility the solution is to produce supranormal numbers of embryos and then hope one survives. But as the example just described illustrates, the technology has run amok. We now live amid an unnatural plague of multiple births and artificial and avoidable brood reduction. In Aldous Huxley's *Brave New World* human reproduction was industrialized in hatchery factories and fetuses reared in bottles. Development was carefully regulated so that each individual could be precisely matched to his or her station in society. Huxley's vision was nightmarish, but the hatchery managers were not incompetent. And their motive was not pecuniary but rather was to maintain a stable social order.

Rapid advances in new reproductive technologies have given new hope to old couples with an even older problem: the inability to have children. Assisted reproduction is founded on the basic principles of parental optimism: produce more incipient offspring than will be reared, select from among these for high-quality embryos, and if necessary reduce the brood. But these natural processes have been taken to unnatural extremes with sometimes frightening outcomes.

An Epidemic of Multiple Births

In vitro fertilization (IVF) is expensive. The cost for a treatment cycle ranges from ten to fifteen thousand dollars, and three to twelve cycles are needed, on average, for each successful pregnancy depending on age; older women take longer. This high cost provides a maximal incentive to make the procedure work. And it works better with more embryos. If two are better than one, then why not three, or four, or six or eight? For a mother-to-be under age thirty a transfer of three embryos nearly trebles the pregnancy rate over that achieved with only a single embryo. This elevated success comes with an elevated risk. Nearly half of those women who become pregnant carry twins or triplets. Assisted reproduction generates an epidemic of multiple births that arises as a direct and avoidable outcome of the technology. Indeed, in the United States today assisted reproduction accounts for fewer than 1% of all births but a third of all twins and 40% of all triplets and higher-order births.

For an assisted reproductive technologies (ART) clinic the bottom line is the take-home baby rate. More embryos mean more babies, and more babies mean more profit. The incentive to transfer too many embryos is great. But more embryos also mean more multiple gestations and more mothers with high-risk and high-cost pregnancies. More fetuses will die before birth, and more babies will be born prematurely, or with low birth weights, or with developmental handicaps. The industry shows no ability to self-regulate, and unscrupulous behavior is fueled by an addiction to ever higher pregnancy rates.

Risks of Multiple Gestation

Human mothers are not designed for multiple gestations, but this is the unfortunate and frequent outcome of IVF. Depending on maternal age and the number of embryos transferred, the rate of multiple pregnancies arising from assisted reproduction is up to fortyfold higher than that of natural pregnancies. This is a dangerous practice for two reasons. First, in normal pregnancy mothers of the most common (dizygotic) type of twins are those women who are phenotypically prepared for the ordeal: tall women with a womb for two and mothers with ample nutritional reserves to pay the heavy costs of lactation. This preparation is absent in assisted reproduction.

And second, mothers and their twin or higher-order babies face myriad and substantial health risks that mothers of singletons do not. Mothers with twins exhibit more first-trimester nausea and vomiting, are at greater risk for preeclampsia (pregnancy-induced high blood pressure, sometimes dangerously high), gestational diabetes, and hemorrhage due to placenta abruptio (placenta tears away from uterine wall) or placenta previa (placenta covers the cervix). Mothers of twins routinely gain forty to eighty pounds during pregnancy to prepare for the heavy costs of lactation. Should a mother not breast-feed, she must work very hard to lose the weight gain. In the United States three-quarters of twin gestations are delivered by caesarian section, which is major abdominal surgery. Even with modern medical care, the risk of maternal mortality is many times higher for twins than for singletons, and the problems do not end once the babies are born. The emotional and physical demands of twins are extreme—the risk of child abuse, for example, was eightfold higher in families with twins than in matched control families with singletons.

The risks to the baby are even more severe. About one in eight pregnancies begins as twins, but only one in eighty to one hundred pregnancies ends in the live birth of twins, and many of the twin conceptions are lost completely. The math is simple—brood reduction occurs often in humans, usually early and out of sight. Some but not all of these losses can be detected with ultrasound examination. Twins are far more likely to be born prematurely, and at low birth weights. Twins suffer more cord accidents at birth, and more physical and mental disabilities, than singletons. The risk of cerebral palsy is tenfold higher in twins than in singletons, and following birth neonatal and infant mortality rates are seven to ten times higher for twins than for singletons. And all these statistics become stratospheric in triplets or higher multiples.

The Ghost in the Machine

The effects of multiple pregnancy, but not necessarily multiple birth, may extend further in ways that we are just beginning to understand. Vanished twins may represent the ghost in the machinery of human reproduction. According to the developmental biologist Charles Boklage, perhaps as many as one in six individuals born as singletons shared a womb with a twin, most of these temporarily (see chapter 8). Given that many of the developmental pathways of the human embryo are decided early in pregnancy, this temporary cohabitation could have far-reaching impacts for the surviving twin.

Biologists have long known that intrauterine companions can have profound effects on development and postnatal behavior, a phenomenon perhaps best studied in mice. In a remarkable series of studies, Fred vom Saal and coworkers revealed a womb-mate effect. Female mouse embryos that are surrounded by males on either side in the womb exhibit masculinized behavior as adults: they are more aggressive and less sexually attractive to the opposite sex. Male mouse embryos that sit between two females become more sexually active as adults and show a greater propensity for infanticidal behavior than embryos with male womb mates. These effects are manifest by the levels of prenatal hormones to which developing embryos are exposed. Individuals situated between two females are exposed to more estrogen; those between two males to more testosterone. And it does not take much of an extra dose to produce these long-lasting effects, just one part per billion of testosterone, even less for estrogen.

What about twins in utero? In cattle, opposite-sex (OS) but not same-sex twin births result in severe developmental abnormalities. The freemartin, who is the female half of the twin pair, is usually rendered sterile by nonfunctional ovaries. In some cases the ovaries are better described as ovotestes, as they contain tissue of both male and female gonads. The bull calf in the twin pair is less severely affected, though often rendered subfertile by abnormal sperm. The exact cause of the freemartin phenomenon is unclear but is likely related to the direct vascular connections that develop between the twins during gestation, resulting in a sharing of blood-borne products (e.g., hormones, stem cells, antigens). Moreover, the freemartin effect can arise even when the male twin dies before birth, the ghost of a twin past. Clearly, an opposite-sex womb mate or two affects offspring quality in mice and cattle. Does it also affect offspring quality in humans? Boklage argues that it does, though in a more subtle fashion than in cattle.

There is growing evidence that sharing a womb with a twin affects the course of development. Opposite-sex twinning in humans affects dental structure and the hearing system. Left-handers (or more properly nonright-handers) are also more common in twin pairs. If Boklage is correct about the true rate of twin gestation, many individuals born as singletons cohabited their mother's womb with a vanished twin. Indeed, many developmental anomalies that are found at higher rates in twin than in singleton births (e.g., cerebral palsy) may be largely or even entirely attributable to the inimical effects of twin gestation. The costs of the blizzard of multiple births associated with assisted reproduction may be far greater than we now realize.

Embryo Reduction

An individual mother faces a fuzzy array of probabilities with multiple embryo transfers, depending on her age, health profile, and bank account. With each round of IVF treatment the most likely outcome is failure (at high financial cost). The next most likely possibility is a singleton pregnancy, with multiple gestation not far behind. Thus, for any individual mother the odds are that she will not have a multiple pregnancy. So why not roll the dice on a multiple embryo transfer, particularly when every throw is the cost of a small car?

The problem is quite different for the reproductive engineer. Probability becomes near certainty the more times we roll the dice. For those who perform multiple embryo transfers a very predictable proportion of treated women will be burdened with multiple gestations, some well beyond the evolutionarily circumscribed limits of human reproduction. Though we celebrate rare successful quadruplet and quintuplet pregnancies, the simple fact is that for every such success there were many disasters. To put it plainly, every multiple gestation is a high-risk pregnancy. A reproductive engineer is putting the mother's own life and the lives of her babies in jeopardy. Even with modern medical science, mothers and especially their babies-to-be face enormous perils, and with every additional fetus the hazards grow exponentially.

The medical engineer has designed a crude, unpalatable solution to the problem: *embryo reduction.* When a multiple embryo transfer results in four, five, six, or more fetuses, the reproductive technician kills the surplus, usually with an injection of potassium chloride in or near the fetal heart, trimming the brood to a manageable number, usually twins. These are avoidable catastrophes. Yet when one scrolls through the medical literature and encounters titles such as "Successful Outcome of Multifetal Reduction in a Pregnancy with 12 Live Fetuses" or "Selective Embryo Reduction in a Sextuplet Pregnancy" one must ask, What were these people thinking? Ever since we shambled off the plains of Africa five million years ago, no human mother has ever given birth to twelve babies. And prior to the latter half of the twentieth century, no human mother had ever carried twelve fetuses. With twelve fetuses there are only two options: nothing is done so that all the fetuses die, and perhaps mother too. Or most of the fetuses are killed so that some will live. Even today twins are an enormous challenge, and only a century ago triplets would have survived intact only very rarely. But sextuplets? Any medical procedure for which a multiple of six fetuses is a possible outcome is clearly corrupt.

Artificial Progeny Choice

Artificial parental optimism creates the opportunity for artificial progeny choice, a practice that has become routine with technological developments that allow human embryos to be cultured outside the human body for three to six days. A series of fertilized embryos can then be inspected for early signs of successful development— how quickly the cells divide, the morphology of structures inside the cell nuclei, whether an embryo shows signs of fragmentation. Those embryos with good prospects for successful implantation are then returned to the prospective mother.

These methods of zygote or blastocyst transfer present a partial solution to the problems of the optimistic medical technologist who transfers multiple embryos in an IVF cycle, creating the epidemic of multiple births. The problem, in a nutshell, is that so many human embryos are of poor quality, particularly in older mothers. That is why spontaneous abortion is so common in humans, and why older mothers, producing ever higher numbers of low-quality embryos, ovulate multiple eggs. We possess built-in systems of screening for embryo quality. This same logic is now extended to programs of assisted reproduction. By screening out low-quality embryos before transfer, medical technologists can considerably reduce the number of embryos transferred. Instead of three, four, five, or more embryos being transferred, with the concomitant risk of triplets, quadruplets, quintuplets, or worse, it should be possible to reduce the number transferred to one or two while still maintaining the same rates of pregnancy (with the possible exception of mothers over age forty using their own eggs, whose prospects for success are very poor even with multiple egg transfers). Surely it is much preferable to transfer fewer embryos at the beginning of pregnancy than to kill supernumeraries when the IVF treatment has been too successful. However, the vexing ethical problem of what to do with the surplus and low-quality embryos not transferred remains.

It is quite clear that in the United States and Canada ART clinics have not been able to self-regulate unethical behavior. The obvious solution is legislative, as occurs, for example, in the United Kingdom and Germany, where the number of embryos transferred in IVF is regulated. But even with legislation there are loopholes, as not all techniques are covered by statute. There is still an open season for the fertility drugs that trigger superovulation and lead to horrific circumstances such as a young mother with a dozen fetuses. Such fertility drugs are a blueprint for reproductive disaster, leading directly to the need to reduce brood size artificially.

Artificial progeny choice has long been a routine component of Western medicine in the form of prenatal screening for birth defects. Screening for Down syndrome is a well-developed science relying on two different methods. One is testing maternal serum in the second trimester of pregnancy for the biochemical markers of trisomy 21 (e.g., high levels of HCG coupled with low levels of alpha fetoprotein). This method is prone to some error, not detecting all cases and registering some false positives. A more accurate albeit invasive method is to sample the amniotic fluid (or chorionic villus) and inspect the cells of fetal origin directly for chromosomal abnormalities. Early detection of Down syndrome allows the parents the option to terminate the pregnancy.

Refining Artificial Progeny Choice

Reproductive technicians have developed a more refined method of progeny choice as part of in vitro fertilization. The method is preimplantation genetic diagnosis (PGD) and relies on a quirk of human reproductive biology. Every cell in the early embryo is capable of developing into a complete adult on its own. When a two-cell embryo splits apart, identical twins result. In other animals, such as many invertebrates, the fate of each cell is predetermined, and the removal of even a single cell has catastrophic and normally fatal consequences. But in humans, IVF technicians can remove one or two cells from an early embryo that usually consists of four to ten cells. DNA from the cell(s) is subjected to molecular analysis using PCR (polymerase chain reaction, which "amplifies" the DNA by making many, many copies and then inspected for the presence of genetic disorders. With PGD a series of embryos can be examined for genetic defects, with only those passing muster being eligible for transfer to the mother-to-be. Though still in its infancy PGD is now used to screen for a variety of genetic disorders including Alzheimer's, Tay-Sachs, and Huntington's diseases, cystic fibrosis, hemophilia A and B, retinoblastoma, retinitis pigmentosa, and muscular dystrophy.

Consider Huntington's disease. It is a degenerative nervous disorder that is incurable and always fatal. It is produced by a single, dominant gene on chromosome 4; thus anyone with a single copy of the gene will develop the disease. Molecular analysis can now reveal whether an individual carries the gene and whether he or she is destined to develop Huntington's in middle age. Carriers of the allele risk passing on the dread disease to their children, as on average half of the children will be afflicted. This genetic time bomb can be defused, however, by preimplantation genetic diagnosis. Embryos prepared by in vitro fertilization can be screened for the Huntington's gene. Surely PGD is an improvement over the trauma of amniocentesis and abortion of the affected fetus in midpregnancy.

Genetic engineers promise an even more effective technique—gene therapy, which would allow the defective gene actually to be repaired in situ. This method holds considerable promise as yet unrealized because of the formidable technical obstacles. It is the nuclear fusion of biotechnology. If and when it works, it will solve many problems, but until then we must reply on the reproductive equivalent of nuclear fission. It is far simpler to check embryos for the presence of a bad gene and not allow them to carry it forward than to try to repair a defective gene later on. Indeed it would probably be prudent to buy stock in biotech companies specializing in PGD rather than gene therapy.

Does Assisted Reproduction Cause Low-Quality Progeny?

Assisted reproduction helps otherwise infertile couples have children. But there is growing evidence that this benefit comes at a cost of reduced progeny quality. As discussed above, the multiple gestation that is far more common in assisted reproduction is itself a risk factor for an adverse pregnancy outcome. But there is reason for further worry. Recent large-scale surveys monitoring the outcome of assisted reproduction report a trio of alarming results. First, singleton babies conceived using assisted reproductive technologies in the United States in 1996 and 1997 faced an almost doubled risk for very low birth weight, though the exact mechanism is unclear. One possibility is that although the pregnancies ended as singleton births, they began as multiple conceptions, and early crowding may have retarded the growth of the surviving baby. As low birth weight elevates the risk for a multitude of adverse health outcomes, this result is cause for concern. Further cause for concern arises from an Australian study showing that assisted reproduction is associated with a doubled risk for multiple birth defects. A Swedish study shows that the risk of neurological problems (e.g., cerebral palsy) in assisted reproduction is double that for natural conceptions even when the effects of low birth weight associated with assisted reproduction were controlled statistically. These results show that assisted reproduction is not a panacea but a flawed tool.

Why does assisted reproduction result in a higher risk of adverse pregnancy outcomes? The answer is unclear, but three possibilities seem likely. First, the mechanical insults associated with assisted reproduction may result in injuries to the developing embryo. The increased risk of monozygotic twinning associated with certain methods of assisted reproduction seems attributable to this cause. Second, assisted reproduction bypasses the built-in physiological mechanisms that regulate pregnancy. Mothers who are phenotypically and/or genotypically unprepared for multiple pregnancy frequently have multiple pregnancy imposed on them, elevating the risk of an adverse outcome. Third, and perhaps most important, the rates of multiple gestation are staggeringly and unnaturally high.

Send in the Clones

Cloning in humans is the holy grail of the new reproductive technologies. Clones are genetic copies of individuals, and monozygotic twins are natural-born clones. Unnatural-born clones are produced in the laboratory by inserting the DNA from an individual into an egg emptied of its nucleus, where most of our DNA is found. This type of genetic engineering is not technically difficult—if only it were that simple. We can send in the clones, of mice and even men (at least to the early embryo stage), but mammalian embryos produced this way are plagued by developmental problems. Most fail to develop at all; those that survive to birth and beyond suffer from further problems—morbid obesity, a variety of behavioral and developmental disorders—what one might call the Homer Simpson phenotype.

We can see the grail clearly, but it for now remains out of reach. The problem is genomic imprinting. In 1989 the evolutionary biologists David Haig and Mark Westoby suggested that you and I are at war with . . . ourselves. We are in fact genetic composites of our parents, and imprinting results in a differential expression of genes from mother and father during development from embryo to adult. Altering the imprinted genes has serious and damaging consequences for the individual, such as neurodevelopmental disorders, abnormalities of growth, and physical malformation (see chapter 5). Assisted reproductions in animals and man are associated with disorders of faulty genomic imprinting, in particular overgrowth of the placenta and embryo. Why is still unclear, but the mechanical insults imposed on the cells during the various procedures seem to distort the patterns of imprinting. Cloning, which involves the removal and replacement of the cell nucleus, is even more intrusive, and dysfunctional imprinting seems to be the norm. This is critical, as early embryo development hinges on the normal functioning of imprinted genes, and the failure rates of cloned embryos in mammals are staggering. Many of the survivors show evidence of seriously abnormal development (born obese, suffer weakness and respiratory failure resulting in mortality shortly after birth, infertility).

Cloning humans by the technology available today would produce a human hecatomb, and only a lucky few would escape alive. And what is just as worrisome is that embryos that appeared normal during gestation and at birth would likely carry imprinting defects that might not be detected until later. The ecstasy that would follow the birth of a cloned baby would often turn to agony as it became obvious that the infant was developmentally abnormal. At some point it will become possible to assess the status of imprinted genes in early embryos for normal function, but that day has not yet arrived. We still do not know how many imprinted genes exist (about fifty are known), where they are located on the chromosomes, and what they do.

Ethical concerns surround all the new reproductive technologies. Each cell removed from an early embryo, and ultimately destroyed in analysis, is capable of developing into a new human being. Embryos that carry a genetic defect are themselves doomed to be destroyed. And there is no reason to restrict PGD to genetic defects. Want a child that is fair-haired and blue-eyed? That is now technologically possible. Worried that your son or daughter (PGD can let you choose) might be left-handed? This too will soon be a possible choice. These possibilities may seem morally repugnant to some, but medical science oozes through the nooks and crannies of the ethical mosaic onto—depending on one's perspective—ever more interesting or terrifying terrain. It is not a question of *whether* these methods will overcome the ethical objections and be brought into practice, but simply when. PGD is technically simple, much more so than gene therapy, though expensive. A PGD analysis currently runs between four thousand and seven thousand dollars, and thus is likely to remain accessible only to the affluent. There will always be someone willing to pay for more and more reproductive choices, and the technology will be developed to meet the need and extract a profit. And PGD works only in conjunction with in vitro fertilization, in which eggs are harvested from women after the administration of fertility drugs and combined with sperm, harvested from men after the administration of pornography, in a petri dish. Aldous Huxley's Brave New World is hurtling toward us.

Parental Optimism and the Law of Unintended Consequences

Under the logic of resource tracking, the excesses of parental optimism are checked by the availability of resources. Food is commonly assumed to be a proxy for resources, but it is in truth a composite of the myriad nutrients required for successful growth, development, and maintenance. A deficiency of just a single nutrient can have dramatic consequences for reproduction, and folic acid is one example. It is critical to cell growth and division, and a shortage is associated with neurological defects at birth such as spina bifida.

The sensible solution to this problem is to fortify the diets of women who are or who wish to be pregnant with extra folic acid. And this works. It averts the birth defects associated with folic acid deficiency. But no good deed goes unpunished, as Clare Booth Luce once observed in a pithy restatement of the law of unintended consequences. Folic acid supplements also elevate the rate of dizygotic twinning, so much so that the net effect is more, not fewer, birth defects. Multiple births are associated with an array of medical problems, chief among them premature birth and low birth weight (see chapter 8). Overall, the cure—folic acid supplementation—is worse than the affliction. But there is hope that the right balance between benefit and cost might be found. Current research is examining dosage effects. Lower doses of folic acid might possibly avert the neural tube defects without grossly elevating the rate of twinning.

The theory of parental optimism also has direct application to the technology of assisted reproduction, which rests on harvesting eggs from mothers-to-be after using hormones that hyperstimulate the ovaries. Doing so bypasses the prescreening phase of progeny choice in humans—the follicle race that is integral to the normal process of ovulation. This follicle competition serves to ensure that only a robust ovum survives, a system that breaks down only when women begin to polyovulate in their late thirties. It is not yet clear whether polyovulation is the cause or symptom of a rising incidence of birth defects in older mothers, but either way a mechanism that increases egg production late in life is likely to end badly. If polyovulation relaxes the screen, then superovulation only makes things worse by recruiting a blizzard of bad embryos alongside whatever good ones remain. And if polyovulation is a symptom of a rising incidence of defective eggs as a woman's supply dwindles late in life, then superovulation only dips deeper into a barrel of rotten eggs. In either case, the natural process of ovulation checks the advance of most defective ova. The technique of assisted reproduction works in the opposite direction, opening the floodgates to defective and normal eggs alike. This is a big part of the reason why assisted reproduction does not work very well for most older women using their own eggs.

The simple fact that ovulation serves as a screening system has largely escaped the attention of the medical sciences, yet it comes as no surprise to evolutionary biologists, who see parallel systems of parental optimism and progeny choice in families throughout nature, from cone-bearing trees to live-bearing sharks to siblicidal eagles and pelicans. As in mice, so too in men . . . and women.

Blame Parents

Technology now extends the reach of the optimistic parent to the Brave New World. Multiple embryos yield multiple births (extra reproductive value), hedge against fertilization failure (insurance), and allow for the screening of birth defects or selection of sex (progeny choice). Stem cells retrieved from the cord blood of newborns are used to treat leukemia in other family members, and in the near future may be used for organ or tissue repair (facilitation). The technology is new, but the motives are not. This is what human parents have done for the last thousand millennia: produce more progeny than may be needed, and trim the excess as conditions warrant. Some offspring serve parents best by dying. Ever since parents acquired the habit of parental care—nurturing incipient progeny beyond the moment of fertilization—they have manipulated offspring for their own interests, often at cross-purposes with their offspring, their closest of genetic relatives. For the last half billion years parents have routinely built more progeny than they were capable of raising. The offspring routinely protest. Some scheme against their parents, some fight internecine battles with brothers and sisters, and some perish early by parental design. Still others cooperate and even serve as agents for their parents' interests. But life's cycle revolves full circle, and the progeny that want too much become the parents that provide too little. The harmony and discord of parent-offspring relations endures.

Chapter 13

Debunking the Family Myth

WHEN A WHITE booby or black eagle nestling pummels its younger brother or sister into a bloody, pulpy mass, is this a dysfunctional family? Is it pathological behavior when a burying beetle cannibalizes its own brood, or a sparrow hawk eats a sibling? Is a male langur's or lion's or mouse's abuse and killing of the infants, cubs, or pups sired by a different male aberrant? In short do we need a battery of animal psychologists (and plant pathologists) to treat seriously deranged parents and offspring, or are these behaviors, spectacular and brutal as they may be, part of the normal behavior repertoire of these species, and explicable with simple Darwinian principles? Quite obviously the latter.

Charles Darwin's remarks regarding the families of lower animals illustrate that even he labored under false expectations. He expected cooperation among family members, just as we still do today. But sometimes conflict is just as important as, or even more than, cooperation. Pregnancy sickness arises in humans, for example, because a mother's hormonal systems have been seized by the embryo she carries, in a dispute over whether a pregnancy should continue. Conflict is even a prerequisite for a successful pregnancy outcome. Remove mother's set of chromosomes from a zygote, replace them with a complete set of dad's, and serious problems arise, as those who attempt to clone humans will soon learn.

Understanding the spectrum of human family relations requires a balanced view of the roles of conflict and cooperation. The assumption that any behavior is necessarily deviant, even those that we find repelling, is both naive and dangerous. The perspective that we can explain human behavior without a Darwinian foundation—still the distorted view of many in the social sciences—is hubris. That said, linking animal to human behavior is no simple task, and having a bit more than a passing familiarity with Geoffrey Parker's work on the theory of parent-offspring conflict and how such conflicts are likely to be resolved (a pro rata compromise most of the time) has not yet helped me in resolving the seemingly endless disputes with my two sons. But there is some comfort in knowing that a bit of greed is good when it comes to parent-offspring relations. That children press the boundaries is natural, expected, . . . and punishable.

We are built from a genetic program that has been field-tested for nearly four billion years. That conflict is still with us—or at least the potential for conflict—speaks to its staying power. Whether potential conflict is actualized—a biologist might describe this as geno-typic (potential) conflict translated into phenotypic (actual) con-flict—often depends on the environment in which a genetic program finds itself. And nothing is more important than the avail-ability of the critical resources—food, space, shelter—required for rearing offspring.

Many biologists find Kurt Vonnegut's novel *Galapagos* hugely en-tertaining because of its remarkable biological accuracy. In it hu-mans are followed from the present (large-brained bipeds with a worldwide distribution) to a future a million years distant, by which time we have evolved into a race of small-brained, flatulent, seal-like creatures found only on the beaches of the Galapagos. I enjoy it even more because Vonnegut concisely captures the essence of so much human conflict, and particularly that within families: "When you get right down to it, food is practically the whole story most of the time."

How true! A poverty of resources is the direct antecedent of child abuse and infanticide in both animals and humans. Limited supply and excess demand animate conflict that has lain dormant within our genetic programming. A blackbird family amid a flush of insects on a warm summer day is a picture of harmony. Warm, well-fed chicks lie contentedly on the nest and do little more than open their mouths when mother returns with a beakful of bugs. But observe that same family after forty-eight hours of cold, wet weather and you see a family in crisis. The brood becomes unsustainable, and every offspring must now look out for itself. While still able, chicks beg frantically for food and attention whenever a parent returns, and often when not. Older, stronger sibs show their weaker counter-parts no quarter, pushing them aside and even trampling them at feeding time. The last hatched within the brood are the first to suffer from neglect, and they rapidly wither and die. The behavior of both parents and offspring, and the potential for harmony or discord, are contingent on circumstance.

How far does this model extend to human behavior? We are just beginning to learn. Humans are quite obviously far more difficult to study than are our feathered counterparts. Whereas a behavioral ecologist can alter the brood size and food levels of bird families experimentally, ethics review committees are far less sympathetic to proposals for similar experiments on human subjects. Thus, to study humans we rely more on observational methods, which give us less control than animal behaviorists customarily enjoy. This simple matter of logistics makes the study of human family relations more challenging.

Child abuse, neglect, and infanticide are human cultural universals. Our behavioral repertoire overlaps with that of organisms as diverse as lemmings and langurs, burying beetles and mourning doves. So how do we rid ourselves of this distasteful evolutionary legacy? We have already tried the approach of societal disapproval through legislation, and it does not work. These behaviors still persist, just less visibly. Any substantive resolution to this problem requires understanding the natural history of conflict in human families and its evolutionary antecedents. A key first step is recognizing that conflict within families is not an aberration.

Families are founded on conflict: parents and offspring are genetically girded for battle. Greed is not only good—to quote the fictional Gordon Gecko—but is sometimes mandatory for a proper functioning of family relations. A nestling bird that does not beg as hard as its brothers or sisters will not get fed. Genes for self-restraint lose the evolutionary game, and unilateral disarmament can be even worse. We see this in the rare aberrancy of molar pregnancy, which occurs when, instead of the normal equal complements of maternal and paternal genes in a developing embryo, a developmental accident results in a double set of the father's chromosomes and a deletion of the mother's chromosomes. Here the normal tension between the genes of mother and those of father is disturbed in favor of dad. The result is catastrophic: an overly aggressive and potentially cancerous growth of the placenta, the organ designed to nourish the embryo. When faced with no opposition, the paternal genes do not hold back but instead run up the score. The result is disastrous for all, as a normal pregnancy relies on a balance between the competing interests.

Parent—offspring relations follow the same path. The normal state of affairs is a standoff between the two, Geoffrey Parker's pro rata compromise. Broods of begging birds are models for the study of this behavior. Nestlings beg, and parents provide. When more than one nestling begs, the volume is turned up, and the displays become more frantic. The parallels to a typical human family are both obvious and instructive. Nestling birds are prone to exaggerate their needs, and parents typically hold something back. Most of the time neither parent nor offspring are working at maximal levels, and neither party gets exactly what it wants. Under this arrangement some level of honesty is required. The selfishness of offspring is tempered though not eliminated by shared genetic interests. The generosity of parents is abridged by an obligation to save something for themselves and for progeny not yet born. The result is a parent-offspring accord that entails offspring asking for more than they need, parents providing less than they can, and neither party being entirely satisfied. And that is as it should be. That such an accord exists is illustrated by interlopers, the brood parasites, who do not play by the same rules. They disturb the equilibrium and in doing so demonstrate both that offspring could get more and that parents could provide more.

We are just beginning to understand the nature of conflict within the human family. We can sketch a rough outline based on the study of animal models. We know that families too large, resources too scarce, congenital defects, birth intervals too close, and families not related closely enough are associated with elevated levels of child abuse, neglect and even infanticide. The role of culture is important. Culture obscures such linkages. It turns what are reflexive actions in birds and other mammals into statistical associations in humans, and hence makes the science of conflict even harder to study. Thus we must learn as much as we can from our nonhuman counterparts, and what we have learned so far is both reassuring and alarming. Reassuring because families are indeed nurturing, protective, and loving, as we expect. And alarming because so often they are not.

Selected References

CHAPTER 1. BLAME PARENTS

Infanticide is shocking to nature:
Hume, David. 1777. *On the Populousness of Ancient Nations.*

Apothetae in ancient Greece and Rome:
Golden, M. 1981. Demography and exposure of girls at Athens. *Phoenix* 35: 330–331.

Brood reduction in blackbirds:
Forbes, S., Glassey, B., Thornton, S., Earle, L. 2001. The secondary adjustment of clutch size in red-winged blackbirds (*Agelaius phoeniceus*). *Behavioral Ecology and Sociobiology* 50: 37–44.
Forbes, S., Glassey, B. 2000. Asymmetric sibling rivalry and nestling growth in red-winged blackbirds. *Behavioral Ecology and Sociobiology* 48: 413–417.
Forbes, S., Grosshans, R., Glassey, B. 2002. Multiple incentives for parental optimism and brood reduction in blackbirds. *Ecology* 83: 2529–2541.

Siblicide in black eagles:
Gargett, V. 1990. *The Black Eagle.* Acorn Books, Randburg, South Africa.

Harpy eagles:
Rettig, N. L. 1978. Breeding behavior of the harpy eagle (*Harpia harpyja*). *Auk* 95: 629–643.

Kin selection:
Hamilton, W. D. 1964. The genetical evolution of social behavior. I. *Journal of Theoretical Biology* 7: 1–16.
Hamilton, W. D. 1964. The genetical evolution of social behavior. II. *Journal of Theoretical Biology* 7: 17–52.

Selfish genes:
Dawkins, R. 1976. *The Selfish Gene.* Oxford University. Press, London.

Parent-offspring conflict:
Trivers, R. L. 1974. Parent-offspring conflict. *American Zoologist* 14: 249–264.

Sibling rivalry:
Mock, D. W., Parker, G. A. 1997. *The Evolution of Sibling Rivalry.* Oxford University Press, Oxford.

Infanticide:

Hausfater, G., Hrdy, S. B. 1984. *Infanticide: Comparative and Evolutionary Perspectives.* Aldine, New York.

Parmigiani, S., vom Saal, F. (eds.) 1994. *Infanticide and Parental Care,* Harwood Academic Publishers, Chur, Switzerland.

CHAPTER 2. THE OPTIMISTIC PARENT

Optimistic parents:

Mock, D. W., Forbes, L. S. 1995. The evolution of parental optimism. *Trends in Ecology and Evolution* 10: 130–134.

Lack clutch size:

Lack, D. 1947. The significance of clutch size. *Ibis* 89: 302–352.

Parental optimism:

Mock, D. W., Forbes, L. S. 1995. The evolution of parental optimism. *Trends in Ecology and Evolution* 10: 130–134.

Mock, D. W., Parker, G. A. 1997. *The Evolution of Sibling Rivalry.* Oxford University Press, Oxford.

Primogeniture and the heir to spare:

Hrdy, S. B., Judge, D. S. 1993. Darwin and the puzzle of primogeniture: An essay of biases in parental investment after death. *Human Nature* 4: 1–45.

CHAPTER 3. WHY PARENTS PLAY FAVORITES

Parental favoritism:

Forbes, L. S., Thornton, S., Glassey, B., Forbes, M., Buckley, N. J. 1997. Why parent birds play favourites. *Nature* 390: 351–352.

Lessels, C. M. 2002. Parentally biased favouritism: Why should parents specialize in caring for different offspring? *Philosophical Transactions of the Royal Society London Series B* 357: 381–403.

Obligate brood reduction:

Anderson, D. J. 1990. Evolution of obligate siblicide in boobies. I. A test of the insurance-egg hypothesis. *American Naturalist* 135: 334–350.

Anderson, D. J. 1990. Evolution of obligate siblicide in boobies. II: Food limitation and parent-offspring conflict. *Evolution* 44: 2069–2082.

Edwards, T. C., Collopy, M. W. 1983. Obligate and facultative brood reduction in eagles: An examination of factors that influence fratricide. *Auk* 100: 630–635.

Stinson, C. H. 1979. On the selective advantage of fratricide in raptors. *Evolution* 33: 1219–1225.

Offspring size vs. number:

Smith, C. C., Fretwell, S. D. 1974. The optimal balance between size and number of offspring. *American Naturalist* 108: 499–506.

Stearns, S. 1992. *The Evolution of Life Histories.* Oxford University Press, Oxford.

Why don't parent birds work harder?

Ydenberg, R. C., Bertram, D. F. 1989. Lack's clutch size hypothesis and brood enlargement studies of colonial seabirds. *Colonial Waterbirds* 12: 134–137.

Parental optimism and insurance:

Anderson, D. J. 1990. Evolution of obligate siblicide in boobies. 1: A test of the insurance-egg hypothesis. *American Naturalist* 135: 334–350.

Forbes, L. S. 1990. Insurance offspring and avian clutch size. *Journal of Theoretical Biology* 147: 345–359.

Forbes, L. S., Lamey, T. C. 1996 Insurance, developmental accidents, and the risks of putting all your eggs in one basket. *Journal of Theoretical Biology* 180: 247–256.

Progeny choice:

Forbes, S., Mock, D. W. 1998. Parental optimism and progeny choice: When is screening for offspring quality affordable? *Journal of Theoretical Biology* 192: 3–14.

Kozlowski, J., Stearns, S. C. 1989. Hypotheses for the production of excess zygotes: Models of bet-hedging and selective abortion. *Evolution* 43: 1369–1377.

Stearns, S. C. 1987. The selection-arena hypothesis. In Stearns, S. C. (ed.), *The Evolution of Sex and Its Consequences*, pp. 337–349. Birkhauser Verlag, Basel.

CHAPTER 4. HOW PARENTS PLAY FAVORITES

Brood reduction in birds:

Amundsen, T., and Slagsvold, T. 1991. Hatching asynchrony: facilitating adaptive or maladaptive brood reduction? *Acta XX Congressus Internationalis Ornithologici*: 1707–1720.

Desmond, A., Moore, J. 1991. *Darwin: The Life of a Tormented Evolutionist.* Norton, New York.

Forbes, L. S., Glassey, B. 2000. Asymmetric sibling rivalry and nestling growth in red-winged blackbirds. *Behavioral Ecology and Sociobiology* 48: 413–417.

Konarzewski, M. 1993. The evolution of clutch size and hatching asynchrony in altricial birds: The effect of environmental variability, egg failure and predation. *Oikos* 67: 97–106.

Lack, D. 1947. The significance of clutch size. *Ibis* 89: 302–352.

Magrath, R. D. 1990. Hatching asynchrony in altricial birds. *Biological Reviews* 65: 587–622.

Mock, D. W. 1994. Brood reduction: Broad sense, narrow sense. *Journal of Avian Biology* 25: 3–7.

Stenning, M. J. 1996. Hatching asynchrony, brood reduction and other rapidly reproducing hypotheses. *Trends in Ecology and Evolution* 11: 243–246.

Stoleson, S. H., Beissinger, S. R. 1997. Hatching asynchrony, brood reduction, and food limitation in a neotropical parrot. *Ecological Monographs* 67: 131–154.

Birth interval and infant mortality:

Hobcraft, J., McDonald, J. W., Rutstein, S. 1983. Child-spacing effects on infant and early child mortality. *Population Index* 49: 585–618.

Maine, B., McNamara, R. 1985. *Birth Spacing and Child Survival.* Center for Population and Family Health, School of Public Health, Faculty of Medicine, Columbia University, NY.

Swenson, I. 1978. Early childhood survivorship related to the subsequent interpregnancy interval and outcome of the subsequent pregnancy. *Journal of Tropical Pediatrics and Environmental Child Health* 24: 103–106.

Trussell, J., Pebley A. R. 1984. The potential impact of changes in fertility on infant, child and maternal mortality. *Studies in Family Planning* 15: 267–280.

Whitworth, A., Stephenson, R. 2002. Birth spacing, sibling rivalry and child mortality in India. *Social Science and Medicine* 55: 2107–2119.

Yerushalmy, J. 1956. Longitudinal studies of pregnancy on the island of Kauai, territory of Hawaii. I. Analysis of previous reproductive history. *American Journal of Obstetrics and Gynecology* 71: 80–96.

Does brood reduction provide relief?

Harper, R. G., Juliano, S. A., Thompson, C. A. 1992. Hatching asynchrony in the house wren, *Troglodytes aedon*: A test of the brood-reduction hypothesis. *Behavioral Ecology* 3: 76–83.

Stoleson, S. H., Beissinger, S. R. 1997. Hatching asynchrony, brood reduction, and food limitation in a neotropical parrot. *Ecological Monographs* 67: 131–154.

Stouffer, P. C., Power, H. W. 1991. An experimental test of the brood-reduction hypothesis in European starlings. *Auk* 108: 519–531.

Brood reduction avoids extra costs:

Gibbons, D. W. 1987. Hatching asynchrony reduces parental investment in the jackdaw. *Journal of Animal Ecology* 56: 403–414.

Bet hedging:

Amundsen, T., Slagsvold, T. 1998. Hatching asynchrony in great tits: A bet-hedging strategy? *Ecology* 79: 295–305.

Laaksonen, T. 2004. Hatching asynchrony as a bet-hedging strategy—an offspring diversity hypothesis? *Oikos* 104: 616–620.

Seger, J., Brockmann, H. J. 1987. What is bet-hedging? *Oxford Surveys in Evolutionary Biology* 4: 182–211.

Sex-biased brood reduction in antechinuses:

Cockburn, A. 1990. Sex ratio variation in marsupials. *Australian Journal of Zoology* 37: 467–479.

Cockburn, A. 1994. Adaptive sex allocation by brood reduction in antechinuses. *Behavioral Ecology and Sociobiology* 35: 53–62.

Cockburn, A., Lee, A. K., Martin, R. W. 1983. Macrogeographic variation in litter size in Antechinus (Marsupialia: Dasyuridae). *Evolution* 37: 86–95.

Davison, M. J., Ward S. J. 1998. Prenatal bias in sex ratios in a marsupial, *Antechinus agilis*. *Proceedings of the Royal Society of London Series B* 265: 2095–2099.

Brood reduction in rabbits:

Drummond, H., Vázquez E., Sánchez-Colón, S., et al. 2000. Competition for milk in the domestic rabbit: Survivors benefit from littermate deaths. *Ethology* 106: 511–526.

Parent plants play favorites:

Janzen, D. H. 1971. Seed predation by animals. *Annual Review of Ecology and Systematics* 2: 465–492.

Lee, T. D. 1988. Patterns of fruit and seed production. In Doust, J. L., Doust, L. L. (eds.). *Plant Reproductive Ecology: Patterns and Strategies*, pp. 179–202. Oxford University Press, New York.

Murneek, A. E. 1954. The embryo and endosperm in relation to fruit development with special reference to the apple, *Malus sylvestris*. *Proceedings of the American Society of Horticultural Science* 64: 573–582.

Stephenson, A. G. 1981. Flower and fruit abortion: Proximate causes and ultimate functions. *Annual Review of Ecology and Systematics* 12: 253–279.

Desperado siblings:

Drummond, H., Rodriguez, C., Vallarino, A., et al. 2003. Desperado siblings: Uncontrollably aggressive junior chicks. *Behavioral Ecology and Sociobiology* 53: 287–296.

Vanishing twins:

Boklage, C. E. 1995. The frequency and survival probability of natural twin conceptions. In Keith, L. G., et al. (eds.), *Multiple Pregnancy*, pp. 41–50. Parthenon Publishing, New York.

Landy, H. J., Keith, L., Keith, D. 1982. The vanishing twin. *Acta Geneticae Medicae et Gemellologiae* 31: 179–194.

Landy, H. J., Nies, B. M. 1995. The vanishing twin. In Keith, L. G., et al. (eds.), *Multiple Pregnancy*, pp. 59–71. Parthenon Publishing, New York.

Siblicide in egrets:

Creighton, J. C., Schnell, G. D. 1996. Proximate control of siblicide in cattle egrets: A test of the food-amount hypothesis. *Behavioral Ecology and Sociobiology* 38: 371–377.

Mock, D. W., Parker, G. A. 1986. Advantages and disadvantages of egret and heron brood reduction. *Evolution* 40: 459–470.

Ploger, B. J., Mock, D. W. 1986. Role of sibling aggression in food distribution to nestling cattle egrets (*Bubulcus ibis*). *Auk* 103: 768–776.

Pending competition:

Forbes, L. S., Ydenberg, R. C. 1992. Sibling rivalry in a variable environment. *Theoretical Population Biology* 41: 135–160.

Stinson, C. 1979. On the selective advantage of fratricide in raptors. *Evolution* 33: 1219–1222.

Avian hatching asynchrony:

Clark, A. B., Wilson, D. S. 1981. Avian breeding adaptations: Hatching asynchrony, brood reduction, and nest failure. *Quarterly Review of Biology* 56: 253 277.

Lack, D. 1947. The significance of clutch size. *Ibis* 89: 302–352.

Magrath, R. D. 1990. Hatching asynchrony in altricial birds. *Biological Reviews of the Cambridge Philosophical Society* 65: 587–622.

Mock, D. W. 1984. Infanticide, siblicide, and avian nestling mortality. In Hausfater, G., Hrdy, S. B. (eds.), *Infanticide: Comparative and Evolutionary Perspectives*, pp. 3–30. Aldine, New York.

Egg size differences in birds:

Christians, J. K. 2002. Avian egg size: Variation within species and inflexibility within individuals. *Biological Reviews of the Cambridge Philosophical Society* 77: 1–26.

Slagsvold, T., et al. 1984. On the adaptive value of intra-clutch egg-size variation in birds. *Auk*, 101: 685–697.

Williams, T. D. 1994. Intraspecific variation in egg composition in birds: Effects on offspring fitness. *Biological Reviews of the Cambridge Philosophical Society* 68: 35–59.

Hormonal titers in birds:

Gil, D., Graves, J., Hazen, N., Wells, A. 1999. Male attractiveness and differential testosterone investment in zebra finch eggs. *Science* 286: 126–128.

Lipar, J. L., Ketterson, E. D. 2000. Maternally derived yolk testosterone enhances the development of the hatching muscle in the red-winged blackbird, *Agelaius phoeniceus*. *Proceedings of the Royal Society of London Series B* 267: 2005–2010.

Schwabl, H. 1996. Maternal testosterone in the avian egg enhances postnatal growth. *Comparative Biochemistry and Physiology* 114A: 271–276.

Schwabl, H. 1999. Developmental changes and among-sibling variation of corticosterone levels in an altricial avian species. *General and Comparative Endocrinology*: 116: 403–408.

Schwabl, H., Mock, D. W., Gieg, J. A. 1997. A hormonal mechanism of parental favouritism. *Nature*, 386: 231.

Sockman, K., Schwabl, H. 2000. Yolk androgens reduce offspring survival. *Proceedings of the Royal Society of London Series B* 267: 1451–1456.

Tinkering with the immune system:

Royle, N. J., Surai, P. F., Hartley, I. R. 2003. The effect of variation in dietary intake on maternal deposition of antioxidants in zebra finch eggs. *Functional Ecology* 17: 472–481.

Unequal food allocation:

Forbes, L. S. 1993. Avian brood reduction and parent-offspring "conflict." *American Naturalist* 142: 82–117.

Mock, D. W. 1987. Siblicide, parent-offspring conflict, and unequal parental investment by egrets and herons. *Behavioral Ecology and Sociobiology* 20: 247–256.

Parker, G. A., Mock, D. W., Lamey, T. C. 1989. How selfish should stronger sibs be? *American Naturalist* 133: 846–868.

Smith, C. C., Fretwell, S. D. 1974. The optimal balance between the number and size of offspring. *American Naturalist* 108: 499–506.

CHAPTER 5. FAMILY CONFLICT

Parent-offspring conflict:

Godfray, H.J.C. 1995. Evolutionary theory of parent-offspring conflict. *Nature* 376: 133–138.

Mock, D. W., Parker, G. A. 1997. *The Evolution of Sibling Rivalry.* Oxford University Press, Oxford.

Trivers, R. 1974 Parent-offspring conflict. *American Zoologist* 14: 249–264.

Wells, J.C.K. 2003. Parent-offspring conflict theory, signaling of need, and weight gain in early life. *Quarterly Review of Biology* 78: 169–202.

Pregnancy, sickness and embryo protection:

Fessler, D.M.T. 2002. Reproductive immunosuppression and diet: An evolutionary perspective on pregnancy sickness and meat consumption. *Current Anthropology* 43: 19–61.

Flaxman, S. M., Sherman, P. W. 2000. Morning sickness: A mechanism for protecting mother and embryo. *Quarterly Review of Biology* 75: 113–148.

Hook, E. B. 1978. Dietary cravings and aversions during pregnancy. *American Journal of Clinical Nutrition* 31: 1355–1362.

Profet, M. 1992. Pregnancy sickness as adaptation: A deterrent to maternal ingestion of teratogens. In Barkow, J., Cosmides, L., Tooby, J. (eds.), *The Adapted Mind*, pp. 327–365. Oxford Univeristy Press, New York.

Profet, M. 1995. *Pregnancy Sickness: Using Your Body's Natural Defenses to Protect Your Baby-to-Be.* Addison-Wesley, New York.

Sherman, P. W., Flaxman, S. M. 2002. Nausea and vomiting of pregnancy in an evolutionary perspective. *American Journal of Obstetrics and Gynecology* 186 (5 Supplement Understanding): S190-S197.

Genetic conflict hypothesis:

Forbes, S. 2002. Pregnancy sickness and embryo quality. *Trends in Ecology & Evolution* 17: 115–120.

Haig, D. 1993. Genetic conflicts in human pregnancy. *Quarterly Review of Biology* 68: 495–532.

Haig, D. 1996. Altercation of generations: Genetic conflicts of pregnancy. *American Journal of Reproductive Immunology* 35: 226–232.

Haig, D. 2004. Evolutionary conflicts in pregnancy and calcium metabolism—a review. *Placenta* 25 (Supplement A): S10–S15.

Genomic imprinting:

Hager, R., Johnstone, R. A. 2003. The genetic basis of family conflict resolution in mice. *Nature* 421: 533–535.

Haig, D. 2000. The kinship theory of genomic imprinting. *Annual Review of Ecology and Systematics* 31: 9–32.

Hurst, L. D., McVean, G. T. Growth effects of uniparental disomies and the conflict theory of genomic imprinting. *Trends in Genetics* 13: 436–443.

Iwasa, Y. 1998. The conflict theory of genomic imprinting: How much can be explained? *Current Topics in Developmental Biology* 40: 255–293.

Weisstein, A. E., Feldman, M. W., Spencer, H. G. 2002. Evolutionary genetic models of the ovarian time bomb hypothesis for the evolution of genomic imprinting. *Genetics* 162: 425–439.

Wilkins, J. F., Haig, D. 2003. What good is genomic imprinting: The function of parent-specific gene expression. *Nature Reviews Genetics* 2003: 359–368.

Molecular evolution of human chorionic gonadotropin:

Maston, G. A., Ruvolo, M. 2002. Chorionic gonadotropin has a recent origin within primates and an evolutionary history of selection. *Molecular Biology Evolution* 19: 320–335.

Frequency of hyperemesis gravidarum:

Fairweather, D.V.I. 1968. Nausea and vomiting in pregnancy. *American Journal of Obstetrics and Gynecology* 102: 135–175.

NVP and early pregnancy loss:

Klebanoff, M. A., Koslowe, P. A., Kaslow, R., Rhoads, G. 1985. Epidemiology of vomiting in early pregnancy. *Obstetrics and Gynecology* 66: 612–616.

Weigel, R. M., Weigel, M. M. 1989. Nausea and vomiting of early pregnancy and pregnancy outcome: A meta-analytical review. *British Journal of Obstetrics and Gynaecology* 96: 1312–1318.

Weigel, M. M., Weigel, R. M. 1989. Nausea and vomiting of early pregnancy and pregnancy outcome. An epidemiological study. *British Journal of Obstetrics and Gynaecology* 96: 1304–1311.

Fetal sex and hyperemesis gravidarum:

Askling, J., Erlandsson, G., Kaijser, M., Akre, O., Ekborn, A. 1999. Sickness in pregnancy and sex of child. *Lancet* 354: 2053.

Basso, O., Olsen, J. 2001. Sex ratio and twinning in women with hyperemesis or pre-eclampsia. *Epidemiology* 12: 747–749.

Hsu, C. D., Witter, F. R. 1993. Fetal sex and severe hyperemesis gravidarum. *International Journal of Gynecology and Obstetrics* 40: 63–64.

Schiff, M. A., Reed, S. D., Daling, J. R. 2004. The sex ratio of pregnancies complimented by hospitalisation for hyperemesis gravidarum. *British Journal of Obstetrics and Gynaecology* 111: 27–30.

Sørenson, H. T., Thulstrup, A. M., Mortensen, J. T., Larsen, H., Pedersen, L. 2000. Hyperemesis gravidarum and sex of child. *Lancet* 355: 407.

Female fetuses produce more HCG:

Bremme, K., Eneroth, P., Nilsson, B. 1982. Hormone levels in amniotic fluid and fetal sex. *Gynecologic and Obstetric Investigations* 14: 245–262.

Masson, G., Anthony, F., Chau, E. 1985. Serum chorionic gonadotrophin (hCG), schwanger-schaftsprotein 1 (SP1), progesterone and oestradiol levels in patients with nausea and vomiting in early pregnancy. *British Journal of Obstetrics and Gynaecology* 92: 211–215.

Twinning and HCG:

Basso, O., Olsen, J. 2001. Sex ratio and twinning in women with hyperemesis or pre-eclampsia. *Epidemiology* 12: 747–749.

Johnson, M. R., Abbas, A., Nicolaides, K. H. 1994. Maternal plasma levels of human chorionic gonadotrophin, oestradiol and progesterone in multifetal pregnancies before and after fetal reduction. *Journal of Endocrinology* 143: 309–312.

Jovanovic, L., Landesman, R., Saxena, B. B. 1977. Screening for twin pregnancy. *Science* 198: 738.

Reuter, A. M., Gaspard, U. J., Deville, et al. 1980. Serum concentrations of human chorionic gonadotropin and its alpha and beta subunits during normal singleton and twin pregnancies. *Clinical Endocrinology* 13: 305–318.

HCG and hyperemesis gravidarum:

Evans, A. J., Li, T. C., Selby, C., Jeftcoate, W. J. 1986. Morning sickness and thyroid function. *British Journal of Obstetrics and Gynaecology* 93: 520–522.

Goodwin, T. M., Montero, M., Mestman, J. H., Pekary, A. E., Hershman, J. M. 1992. The role of chorionic gonadotropin in transient hyperthyroidism of hyperemesis gravidarum. *Journal of Clinical Endocrinology and Metabolism* 75: 1333–1337.

Hershmann, J. M. 2004. Physiological and pathological aspects of the effect of human chorionic gonadotropin on the thyroid. *Best Practice & Research Clinical Endocrinology & Metabolism* 18: 249–265.

Kauppila, A., Huhtaniemi, I., Ylikorkala, O. 1979. Raised serum human chorionic gonadotrophin concentrations in hyperemesis gravidarum. *British Medical Journal* 1: 1670–1671.

Kennedy, R. L., Darne, J. 1991. The role of HCG in regulation of the thyroid gland in normal and abnormal pregnancy. *Obstetrics and Gynecology* 78: 298–307.

Rodien, P., Jordan, N., Lefevre, A., et al. 2004. Abnormal stimulation of the thyrotrophin receptor during gestation. *Human Reproduction Update* 10: 95–105.

Schoeneck, F. J. 1942. Gonadotropic hormone concentration in emesis gravidarum. *American Journal of Obstetrics and Gynecology* 43: 308.

HCG is imprinted:

de Grood, N., Goshen, R., Rachmilewitz, J., et al. 1993. Genomic imprinting and b-chorionic gonadotropin. *Prenatal Diagnosis* 13: 1159–1160.

Haig, D. 1993. Genomic imprinting, human chorionic gonadotropin, and triploidy. *Prenatal Diagnosis* 13: 151.

Molar pregnancy:

Altieri, A., Franceschi, S., Ferlay, J., et al. 2003. Epidemiology and aetiology of gestational trophoblastic diseases. *Lancet Oncology* 4: 670–678.

Dawood, M. Y., Saxena, B. B., Landesman, R. 1977. Human chorionic gonadotropin and its subunits in hydatidiform mole and choriocarcinoma. *Obstetrics and Gynecology* 50: 172–181.

Fairweather, DVI 1968. Nausea and vomiting in pregnancy. *American Journal of Obstetrics and Gynecology* 102: 135–175.

Fisher, R. A., Hodges, M. D. 2003. Genomic imprinting in gestational trophoblastic disease—a review. *Placenta* 24 (Supplement A) 17: S111–S118.

Hershman, J. M. 1991. Trophoblastic tumors and hyperthyroidism. In Braverman, L. E., Utiger, R. D. (eds.), *The Thyroid*, pp. 705–709. J. B. Lippincott, Philadelphia.

Imprinted genes and human disease:

Goshen, R., et al. 1994. The role of genomic imprinting in implantation. *Fertility and Sterility* 62: 903–910.

Haig, D., Graham, C. 1991. Genomic imprinting and the strange case of the insulin-like growth factor-II receptor. *Cell* 64: 1045–1046.

Haig, D., Wharton, R. 2003. Prader-Willi syndrome and the evolution of human childhood. *American Journal of Human Biology* 15: 320–329.

Hall, J. G. 1999. Human diseases and genomic imprinting. *American Journal of Human Biology: Results and Problems in Cell Differentiation* 25: 119–132.

Lighten, A. D., Hardy, K., Winston, R.M.L., Moore, G. E. 1997. IGF2 is parentally imprinted in human preimplantation embryos. *Nature Genetics* 15: 122–123.

Morison, I. M., Reeve, A. E. 1998. A catalogue of imprinted genes and parent-of-origin effects in humans and animals. *Human Molecular Genetics* 7: 1599–1609.

Polychronakos, C., Kukuvitis, A. 2002. Parental genomic imprinting in endocinopathies. *European Journal of Endocrinology* 147: 561–569.

Spencer, H. G. 2000. Population genetics and evolution of genomic imprinting. *Annual Review of Genetics* 34: 457–477.

Turner's syndrome:

Kesler, S. R., Blasey, C. M., Brown, W. E., et al. 2003. Effects of X-monosomy and X-linked imprinting on superior temporal gyrus morphology in Turner syndrome. *Biological Psychiatry* 54: 636–646.

Ranke, M. B., Saenger, P. 2001. Turner's syndrome. *Lancet* 358: 309–314.

Skuse, D. H., James, R. S., Bishop, D.V.M., et al. 1997. Evidence from Turner's syndrome of an imprinted X-linked locus affecting cognitive function. *Nature* 387: 705–708.

Preeclampsia:

Haig, D. 1993. Genetic conflicts in human pregnancy. *Quarterly Review of Biology* 68: 495–532.

Lachmeijer, A.M.A., Dekker, G. A., Pals, G., et al. 2002. Searching for preeclampsia genes: The current position. *European Journal of Obstetrics and Gynecology* 105: 94–113.

Marshall, Graves, J. A. 1998. Genomic imprinting, development and disease—is pre-eclampsia caused by a maternally imprinted gene? *Reproduction, Fertility and Development* 10: 23–29.

Pipkin, F. B., Rubin, P. C. 1994. Pre-eclampsia—'the disease of theories.' *British Medical Bulletin* 50: 381–396.

Redman, CWG, Sargent, I. L. 2003. Pre-eclampsia, the placenta and the maternal systemic inflammatory response—a review. *Placenta* (Supplement A) 24: S21–S27.

Robillard, P.-Y. 2002. Interest in preeclampsia for researchers in reproduction. *Journal of Reproductive Immunology* 53: 279–287.

CHAPTER 6. SELFISHNESS UNCONSTRAINED

Brood parasitic birds:

Davies, N. 2000. *Cuckoos, Cowbirds and Other Cheats*. Princeton University Press, Princeton, N.J.

Johnsgard, P. A. 1997. *The Avian Brood Parasites: Deception at the Nest*. Oxford University Press, New York.

Kilner, R. M. How selfish is a cowbird nestling? *Animal Behavior* 66: 569–296.

Ortega, C. 1998. *Cowbirds and Other Brood Parasites*. University of Arizona Press, Tucson.

Rothstein, S. I., Robinson, S. K. (eds.). 1999. *Parasitic Birds and Their Hosts: Studies in Coevolution*. Oxford University Press, New York.

Host-parasite coevolution:

Briskie, J. V., Sealy, S. G., Hobson, K. A. 1992. Behavioral defenses against avian brood parasitism in sympatric and allopatric host populations. *Evolution* 46: 334–340.

Davies, N. B., de L. Brooke, M. 1988. Cuckoos versus reed warblers: Adaptations and counteradaptations. *Evolutionary Ecology* 4: 35–42.

Lyon, B. E. 2003. Egg recognition and counting reduce costs of avian conspecific brood parasitism. *Nature* 422: 495–499.

Rothstein, S. 1990. A model system of coevolution: Avian brood parasitism. *Annual Review of Ecology and Systematics* 58: 207–224.

Sealy, S. G. 1996. Evolution of host defences against brood parasitism: Implications of puncture-ejection by a small passerine. *Auk* 113: 346–355.

Clutch size in cowbirds:

Alderson, G. W., Gibbs, H. L., Sealy, S. G. 1999. Determining the reproductive behaviour of individual brown-headed cowbirds using microsatellite DNA markers. *Animal Behavior* 58: 895–905.

Cowbirds as predators:

Arcese, P., Smith, J. N., Hatch, M. I. 1996. Nest predation by cowbirds and its consequences for passerine demography. *Proceedings of the National Academy of Sciences* 93: 4608–4611.

Avian invasion:

Cruz, A., et al. 1985. The shiny cowbird: A brood parasite expanding its range in the Caribbean region. In Buckley, P. A., et al. (eds.), *Neotropical Ornithology*, Ornithological Monographs 36: 607–620. American Ornithologists Union, Washington, D.C.

Jaramillo, A., Burke, P. 1999. *New World Blackbirds*. Princeton University Press, Princeton, NJ.

Brood parasitism in ducks:

Sorenson, M. 1991. The functional significance of parasitic egg laying and typical nesting in redhead ducks: An analysis of individual behaviour. *Animal Behavior* 42: 771–796.

Sorenson, M. 1997. Effects of intra- and interspecific brood parasitism on a precocial host, the canvasback, *Aythya valisineria*. *Behavioral Ecology* 8: 153–161.

Brood parasitic catfish:

Sato, T. 1986. A brood parasitic catfish of mouth brooding cichlid fishes in Lake Tanganyika. *Nature* 323: 58–59.

Slave-making ants:

Hölldobler, B., Wilson, E. O. 1990. *The Ants*. Belknap Press of Harvard University Press, Cambridge, MA.

Egg-stealing darters:

Porter, B. A., Fiumera, A. C., Avise, J. C. 2002. Egg mimicry and alloparental care: Two mate-attracting tactics by which nesting striped darter (*Etheostoma virgatum*) males enhance reproductive success. *Behavioral Ecology and Sociobiology* 51: 350–359.

Avian mafia:

Soler, M., Soler, J. J., Martinez, J. G., Moller, A. P. 1995. Magpie host manipulation by great spotted cuckoos: Evidence for an avian mafia? *Evolution* 49: 770–775.

CHAPTER 7. SCREENING FOR OFFSPRING QUALITY

Theory of progeny choice:

Buchholz, J. T. 1922. Developmental selection in vascular plants. *Botanical Gazette* 73: 249–286.

Forbes, L. S., Mock, D. W. 1998. Parental optimism and progeny choice: When is screening for offspring quality affordable? *Journal of Theoretical Biology* 192: 3–14.

Kozlowski, J., Stearns, S. C. 1989. Hypotheses for the production of excess zygotes: Models of bet-hedging and selective abortion. *Evolution* 43: 1369–1377.

Maternal screening in humans:

Stein, Z., et al. 1986. Attrition of trisomies as a maternal screening device. *Lancet* 327: 944–947.

Birth defects in humans:

Morton, C. C., Miron, P. 1999. Cytogenetics in reproduction. In Yen, S.S.C., et al. (eds.), *Reproductive Endocrinology*, 4th ed., pp. 323–344. W. B. Saunders, Philadelphia.

Rimoin, D. L., et al. (eds.). 1996. *Emery and Rimoin's Principles and Practice of Medical Genetics*, vol. 1, 3rd ed. Churchill Livingstone, New York.

Down syndrome:

Cohen, W. I. 1999. Health care guidelines for individuals with Down syndrome: 1999 revision. *Down Syndrome Quarterly*. 4: 1–16.

Freeman, S. B., Yang, Q. H., Allran, K., et al. 2000. Women with a reduced ovarian complement may have an increased risk for a child with Down syndrome. *American Journal of Human Genetics* 66: 1680–1683.

Hook, E. B., Mutton, D. E., Ide, R., et al. 1995. The natural history of Down syndrome conceptuses diagnosed prenatally that are not electively terminated. *American Journal of Human Genetics* 57: 875–881.

Jacobs, P. A., Hassold, T. J. 1995. The origin of numerical chromosome abnormalities. *Advances in Genetics* 33: 101–133.

Nicolaidis, P., Petersen, M. B. 1998. Origin and mechanisms of non-disjunction in human autosomal trisomies. *Human Reproduction* 13: 313–319.

Penrose, L. S. 1933. The relative effects of paternal and maternal age in mongolism. *Journal of Genetics* 27: 219–224.

Reeves, R. H., Baxter, L. L., Richtsmeier, J. T. 2001. Too much of a good thing: Mechanisms of gene action in Down syndrome. *Trends in Genetics* 17: 83–88.

Physiological aging and aneuploidy:

Brook, J. D., Gosden, R. G., Chandley, A. C. 1984. Maternal ageing and aneuploid embryos: Evidence from the mouse that biological and not chronological age is the important influence. *Human Genetics* 66: 41–45.

Maternal age and embryo quality:

Armstrong, D. T. 2001. Effects of maternal age on oocyte developmental competence. *Theriogenology* 55: 1303–1322.

Janny, L., Menezo, Y.J.R. 1996. Maternal age effect on early human embryonic development and blastocyst formation. *Molecular Reproduction and Development* 45: 31–37.

te Velde, E. R., Scheffer, G. J., Dorland, M., et al. 1998. Developmental and endocrine aspects of normal ovarian aging. *Molecular and Cellular Endocrinology* 145: 67–73.

Ziebe, S., Loft, A., Petersen, J. H., et al. 2001. Embryo quality and developmental potential is compromised by age. *Acta Obstetricia et Gynecologica Scandinavica* 80: 169–174.

Oocyte selection hypothesis:

Zheng, C.-J., Byers, B. 1992. Oocyte selection: A new model for the maternal-age dependence of Down syndrome. *Human Genetics* 90: 1–6.

Proportion of chromosomally abnormal conceptions:

Bahce, M., et al. 1999. Preimplantation genetic diagnosis of aneuploidy: Were we looking at the wrong chromosomes? *Journal of Assisted Reproduction and Genetics* 16: 176–181.

Boue, J., Boue, A., Lazar, P. 1975. Retrospective and prospective epidemiological studies of 1500 karyotyped spontaneous human abortuses. *Teratology* 12: 11–26.

Hertig, A. T., Rock, J., Adams, E. C., Menkin, M. C. 1959. Thirty-four fertilized human ova, good, bad and indifferent, recovered from 210 women of known fertility: A study of biologic wastage in early human pregnancy. *Pediatrics* 23L: 202–211.

Kajii, T., Ohama, K., Mikamo, K. 1978. Anatomic and chromosomal anomalies in 944 induced abortuses. *Human Genetics* 43: 247–258.
Schmidt-Sarosi, C., et al. 1998. Chromosomal analysis of early fetal losses in relation to transvaginal ultrasonographic detection of fetal heart motion after infertility. *Fertility and Sterility* 69: 274–277.

Aneuploidy in humans:

Bond, D. J., Chandley A. C. 1983. *Aneuploidy.* Oxford Monographs on Medical Genetics No. 11. Oxford University Press, New York.
Hassold, T. 1986. Chromosome abnormalities in human reproductive wastage. *Trends in Genetics* 2: 105–110.
Hassold, T., Hunt, P. 2001. To err (meiotically) is human: The genesis of human aneuploidy. *Nature Reviews Genetics* 2: 280–291.
Hassold, T. J., Jacobs, P. A. 1984. Trisomy in man. *Annual Review of Genetics* 18: 69–97.

Uniparental disomy and trisomy rescue:

Kotzot, D. 2004. Advanced parental age in maternal uniparental disomy (UPD): Implications for the mechanism of formation. *European Journal of Human Genetics* 12: 343–346.
Ledbetter, D. H., Engel, E. 1995. Uniparental disomy in humans: Development of an imprinting map and its implications for prenatal diagnosis. *Human Molecular Genetics* 4 Special No: 1757–64.

Spontaneous, abortion in humans:

Edmonds, D. K., Lindsay, K. S., Miller J. F., et al. 1982. Early embryonic mortality in women. *Fertility and Sterility* 38: 447–453.
Forbes, L. S. 1997. The evolutionary biology of spontaneous abortion in humans. *Trends in Ecology & Evolution* 12: 446–450.
Goddijn, M., Leschot, N. J. 2000. Genetic aspects of miscarriage. *Baillière's Clinical Obstetrics and Gynaecology* 14: 855–865.
Hassold, T., Chen, N., Funkhouser, J., et al. 1980. A cytogenetic study of 1000 spontaneous abortions. *Annals of Human Genetics* 44: 151–165.
Lummaa, V., Haukioja, E., Lemmetyinen, R. 1999. Does polyovulation counterbalance a high abortion rate in humans? *Journal of Evolutionary Biology* 12: 806–808.
Miller, J. F., et al. 1980. Fetal loss after implantation. *Lancet* 316: 554–556.
Roberts, C. J., Lowe, C. R. 1975. Where have all the conceptions gone? *Lancet* 305: 498–499.

Turner's syndrome and imprinting:

Iwasa, Y., Pomiankowski, A. 2001. The evolution of X-linked genomic imprinting. *Genetics* 158: 1801–1809.
Skruse, D. H., James, R. S., Bishop D. V., et al. 1997. Evidence from Turner's syndrome of an imprinted X-linked locus affecting cognitive function. *Nature* 387: 705–708.

CHAPTER 8. WHY TWINS?

Biology of twinning:

Bulmer, M. G. 1970. *The Biology of Twinning in Man.* Oxford University Press, Oxford.

Busjahn, A., Knoblauch, H., Faulhaber, H. D., et al. 2000. A region on chromosome 3 is linked to dizygotic twinning. *Nature Genetics* 26: 398–399.

Gilfillan, C. P., Robertson, D. M., Burger H. G., et al. 1996. The control of ovulation in mothers of dizygotic twins. *Journal of Clinical Endocrinology and Metabolism* 81: 1557–1562.

Hall, J. G. 2003. Twinning. *Lancet* 362: 735–743.

Keith, L. G., et al. (eds.). 1995. *Multiple Pregnancy: Epidemiology, Gestation and Perinatal Outcome.* Parthenon Publishing, New York.

Lambalk, C. B., De Koning, C. H., Braat DDM. 1998. The endocrinology of dizygotic twinning in the human. *Molecular and Cellular Endocrinology* 145: 97–102.

Twinning in IVF:

Schieve, L. A., Peterson, H. B., Meikle, S. F., et al. 1999. Live-birth rates and multiple birth risk using in vitro fertilization. *Journal of the American Medical Association* 282: 1832–1838.

Insurance and replacement in humans:

Montgomery, M. R., Cohen, B. (eds.) 1998. *From Death to Birth: Mortality Decline and Reproductive Change.* National Academy Press, Washington, DC.

Preston, S. H. (ed.). 1978. *The Effect of Infant and Child Mortality on Fertility.* Academic Press, New York.

Fault-tolerant design:

NASA., 1995. Fault tolerant design. *NASA Preferred Reliability Practices*, Practice No. PD-ED-1246.

NASA., 1995. Redundancy switching analysis. *NASA Preferred Reliability Practices*, Practice No. PD-AP-1315.

Lack clutch size:

Lack, D. 1947. The significance of clutch size. *Ibis* 89: 302–352.

Stearns, S. C. 1992. *The Evolution of Life Histories.* Oxford University Press, Oxford.

Paternal transmission of twinning:

Boklage, C. E. 2004. The biology of human twinning: A needed change of perspective. In Keith, L. G., Blickstein, I. (eds.), *Multiple Pregnancy*, 2nd ed., Parthenon Publishing, New York.

Golubovsky, M. D. 2002. Paternal familial twinning: Hypotheses and genetic/medical implications. *Twin Research* 5: 75–86.

Golubovsky, M. D. 2003. Postzygotic diploidization of triploids as a source of unusual cases of mosaicism, chimerism and twinning. *Human Reproduction* 18: 236–242.

McFadden, D. E., Jiang, R., Langlois, S., et al. 2002. Dispermy—origin of diandric triploidy. *Human Reproduction* 17: 3037–3038.

St. Clair, J. B., Golubovsky, M. D. 2002. Paternally derived twinning: A two-century examination of records of one Scottish name. *Twin Research* 5: 294–307.

Rapid evolution of genes involved in fertilization:

Civetta, A., Singh, R. S. 1998. Sex-related genes, directional selection, and speciation. *Molecular Biology and Evolution* 15: 901–909.

Crackower, M. A. 2003. Essential role of Fkbp6 in male fertility and homologous chromosome pairing in meiosis. *Science* 300: 1291–1295.

Glassey, B., Civetta, A. 2004. Positive selection at reproductive ADAM genes with potential intercellular binding activity. *Molecular Biology and Evolution* 21: 851–859.

Parker, G. A., Partridge, L. 1998. Sexual conflict and speciation. *Philosophical Transaction Royal Society of London Series B* 353: 261–274.

Individual optimization in great tits:

Pettifor, R. A., et al. 1988. Variation in clutch-size in great tits: Evidence for the individual optimisation hypothesis. *Nature* 336: 160–162.

Insurance offspring in birds:

Anderson, D. J. 1990. Evolution of obligate siblicide in boobies. I: A test of the insurance egg hypothesis. *American Naturalist* 135: 334–350.

Cash, K., Evans, R. M. 1986. Brood reduction in the American white pelican, *Pelecanus erythrorhynchos*. *Behavioral Ecology and Sociobiology* 18: 413–418.

Evans, R. M. 1996. Hatching asynchrony and survival of insurance offspring in an obligate brood reducing species, the American white pelican. *Behavioral Ecology and Sociobiology* 39: 203–209.

Forbes, L. S. 1990. Insurance offspring and the evolution of avian clutch size. *Journal of Theoretical Biology* 147: 345–359.

Forbes, S., Thornton, S., Glassey, B., Forbes, M., Buckley, N. 1997. Why parent birds play favourites. *Nature* 390: 351–352.

Lundberg, S. 1985. The importance of egg hatchability and nest predation in clutch size evolution in altricial birds. *Oikos* 45: 110–117.

Fitness of twinning:

Gabler, S., Voland, E. 1994. Fitness of twinning. *Human Biology* 66: 699–713.

Haukioja, E., Lemmetyinen, R., Pikkola, M. 1989. Why are twins so rare in *Homo sapiens*? *American Naturalist* 133: 572–577.

Hella, S., Lummaa, V., Jokela, J. 2004. Selection for increased brood size in historical human populations. *Evolution* 58: 430–436.

Lummaa, V., Haukioja, E., Lemmetyinen, R., et al. 1998. Natural selection on human twinning. *Nature* 394: 533–534.

Lummaa, V., Jokela, J., Haukioja, E. 2001. Gender difference in benefits of twinning in pre-industrial humans: Boys did not pay. *Journal of Animal Ecology* 70: 739–746.

Sear, R., Shanley, D., McGregor I. A., et al. 2001. The fitness of twin mothers: Evidence from rural Gambia. *Journal of Evolutionary Biology* 14: 433–443.

Twinning and maternal height and weight:

Brown, J. E., Schloesser, P. T. 1990. Prepregnancy weight status, prenatal weight gain, and the outcome of term twin gestations. *American Journal of Obstetrics and Gynecology* 162: 182–186.

Corney, G., Seedburgh, D., Thompson, B., et al. 1979. Maternal height and twinning. *Annals of Human Genetics* 43: 55–59.

MacGillivray, I., Campbell, D. M. 1978. The physical characteristics and adaptations of women with twin pregnancies. In Allen, G., et al. (eds.), *Twin Research: Clinical Studies*, pp. 81–86. Alan R Liss, New York.

Wyshak, G. 1981. Reproductive and menstrual characteristics of mothers of multiple births and mothers of singletons only: A discriminant analysis. In Gedda, L., Parisi, P., Nance, W. E. (eds.), *Twin Research 3: Twin Biology and Multiple Pregnancy*, pp. 95–105. Alan, R. Liss, New York.

Natural selection on the frequency of twins in Finland:

Lummaa, V., Haukioja, E., Lemmetyinen, R., Pikkola, M. 1998. Natural selection on human twinning. *Nature* 394: 533–534.

Monozygotic twinning in malnourished women:

Bulmer, M. G. 1959. Twinning rate in Europe during the war. *British Medical Journal* 1: 29–30.

Eriksson, A. W., Bressers, W.M.A., Kostense, P. J., et al. 1988. Twinning rate in Scandinavia, Germany and the Netherlands during years of privation. *Acta Genetica Medicae at Gemellologiae* 37: 277–297.

Lactation and twins:

Worthington-Roberts, B., Williams S. R. (eds.). 1989. *Nutrition in Pregnancy and Lactation.* Times Mirror/Mosby, St. Louis.

Twinning and child abuse:

Bryan, E. 2003. The impact of mulitiple preterm births on the family. *British Journal of Obstetrics and Gynaecology* 110 (Supplement 20): 24–28.

Groothuis, J. R., Altemeier, W. A., Robarge, J. P., et al. 1982. Increased child abuse in families with twins. *Pediatrics* 70: 769–773.

Prenatal weight gain and infant birth weight:

Abrams, B. F., Laros, R. K. 1986. Prepregnancy weight, weight gain, and birth weight. *American Journal of Obstetrics and Gynecology* 154: 503–509.

Brown, J. E., Schloesser, P. T. 1990. Prepregnancy weight status, prenatal weight gain, and the outcome of term twin gestations. *American Journal of Obstetrics and Gynecology* 162: 182–186.

Twinning as an atavism:

Anderson, W.J.R. 1956. Stillbirth and neonatal mortality in twin pregnancy. *Journal of Obstetrics and Gynaecology of British Empire* 63: 205–215.

Intrauterine volume in twins versus singletons:

Redford, D.H.A. 1982. Uterine growth in twin pregnancy by measurements of total intrauterine volume. *Acta Genetica Medicae at Gemellologiae* 31: 145–148.

Prenatal growth in twins:

McKeown, T., Record, R. G. 1953. The influence of placental size of foetal growth in man, with special reference to multiple pregnancy. *Journal of Endocrinology* 9: 418–426.

Worthington-Roberts, B. 1988. Weight gain patterns in twin pregnancies with desirable outcomes. *Clinical Nutrition* 7: 191–196.

Maternal height and optimum birth outcomes in twins:

Pederson, A. L., Worthington-Roberts, B., Hickok, D. E. 1989. Weight gain patterns during twin gestation. *Journal of the American Dietetic Association* 89: 642–646.

Women with multiples live closer to the energetic edge:

Gluckman, P. D., Breier, B. H., Oliver, M., et al. 1990. Fetal growth in late gestation—a constrained pattern of growth. *Acta Paediatrica Scandinavica* (Supplement) 367: 105–110.

Weiss, W., Jackson, E. C. 1969. Maternal factors affecting birth weight. *Proceedings of the Pan American Health Organization Advisory Committee on Medical Research* 1: 54–59.

Human twinning and insurance:

Anderson, D. J. 1990. On the evolution of human brood size. *Evolution* 44: 438–440.

Ball, H. L., Hill, C. M. 1998. Insurance ovulation, embryo mortality and twinning. *Journal of Biosocial Science* 31: 245–255.

Down syndrome and dizygotic twinning:

Hamerton, J. L., Briggs, S. M., Giannelli, F., Carter, C. O. 1961. Chromosome studies in detection of parents with high risk of a second child with Down's syndrome. *Lancet* 278: 788–791.

Liberfarb, R. M., Atkins, L., Holmes, L. B. 1978. Down syndrome in two of three triplets. *Clinical Genetics* 14: 261–264.

Lubs, H. A., Ruddle, F. H. 1970. Chromosomal abnormalities in the human population: Estimation of rates based on New Haven newborn study. *Science* 169: 495–497.

McDonald, A. D. 1964. Mongolism in twins. *Journal of Medical Genetics* 1: 39–41.

Sharav, T. 1991. Aging gametes in relation to incidence, gender, and twinning in Down syndrome. *American Journal of Medical Genetics* 39: 116–118.

Oocyte selection:

Zheng, C. J., Byers, B. 1992. Oocyte selection: A new model for the maternal-age dependence of Down syndrome. *Human Genetics* 90: 1–6.

The 90% rule of obligate brood reduction:

Simmons, R. E. 1988. Offspring quality and the evolution of Cainism. *Ibis* 130: 339–357.

Obligate brood reduction in humans:

Boklage, C. E. 1990. Survival probability of human conceptions from fertilization to term. *International Journal of Fertility* 35: 75–94.

Boklage, C. E. 1995. The frequency and survival probability of natural twin conceptions. In Keith, L. G., et al. (eds.), *Multiple Pregnancy: Epidemiology, Gestation and Perinatal Outcome*, pp. 41–50. Parthenon Publishing, New York.

Tong, S., Meagher, S., Vollenhoven, B. 2002. Dizygotic twin survival in early pregnancy. *Nature* 416: 142.

CHAPTER 9. FATAL SIBLING RIVALRY

Selfish genes:

Dawkins, R. 1976. *The Selfish Gene.* Oxford University Press, London.

Kin selection and Hamilton's rule:

Hamilton, W. D. 1964. The genetical evolution of social behavior. *Journal of Theoretical Biology* 7: 1–16.

Mock, D. W., Parker, G. A. 1997. *The Evolution of Sibling Rivalry.* Oxford University Press, New York.

Insurance offspring in birds:

Anderson, D. J. 1990. Evolution of obligate siblicide in boobies. I: A test of the insurance egg hypothesis. *American Naturalist* 135: 334–350.

Cash, K., Evans, R. M. 1986. Brood reduction in the American white pelican (*Pelecanus erythrorhynchos*). *Behavioral Ecology and Sociobiology* 18: 413–418.

Forbes, L. S. 1990. Insurance offspring and the evolution of avian clutch size. *Journal of Theoretical Biology* 147: 2345–2359.

Bald eagle siblicide:

Bortolotti, G. R. 1986. Influence of sibling competition on nestling sex ratios of sexually dimorphic birds. *American Naturalist* 127: 495–507.

Facultative siblicide:

Forbes, L. S., Mock, D. W. 1994. Proximate and ultimate determinants of avian brood reduction. In Parmigiani, S., vom Saal, F. (eds.), *Infanticide and Parental Care*, pp. 237–256. Harwood Academic Publishers, New York.

Mock, D. W. 1984. Infanticide, siblicide, and avian nestling mortality. In Hausfater, G., Hrdy, S. B. (eds.), *Infanticide: Comparative and Evolutionary Perspectives*, pp. 3–30. Aldine, New York.

Obligate siblicide in birds:

Forbes, L. S., Ydenberg, R. C. 1992. Sibling rivalry in a variable environment. *Theoretical Population Biology* 41: 335–360.

Godfray, H.C.J., Harper, A. B. 1990. The evolution of brood reduction by siblicide in birds. *Journal of Theoretical Biology* 145: 163–175.

Stinson, C. H. 1979. On the selective advantage of fratricide in raptors. *Evolution* 33: 1219–1225.

Sibling aggression in blue-footed boobies:

Drummond, H., Garcia Chavelas, C. 1989. Food shortage influences sibling aggression in the blue-footed booby. *Animal Behavior* 37: 806–819.

Cannibalism in sand tiger sharks:

Gilmore, R. J., Dodrill, J. W., Linley, P. A. 1983. Reproduction and development of the sand tiger shark, *Odontapsis taurus (Rafinesque)*. *Fishery Bulletin* 81: 201–225.

Hyena siblicide:

Frank, L. G., Glickman, S. E., Licht, P. 1991. Fatal sibling aggression, precocial development, and androgens in neonatal spotted hyenas. *Science* 252: 702–704.

Smale, L., Holekamp, K. E., White, P. A. 1999. Siblicide revisited in the spotted hyena: Does it conform to obligate or facultative models? *Animal Behavior* 58:545–551.

Pitcher plant flies:

Forsyth, A. B., Robertson, R. J. 1975. K reproductive strategy and larval behavior of the pitcher plant sarcophagid fly, *Blaesoxipha fletcheri*. *Canadian Journal of Zoology* 53: 174–179.

MEDEA alleles in flour beetles:

Beeman, R. W., Friesen, K. S., Denell, R. E. 1992. Maternal-effect selfish genes in flour beetles. *Science* 256: 89–92.

Monozygotic versus dizgyotic twinning in humans:

Bulmer, M. G. 1972. *The Biology of Twinning in Man*. Clarendon Press, Oxford.

MacGillivray, I., Campbell, D., Thompson, B. (eds.). 1988. *Twinning and Twins*. John Wiley & Sons, New York.

Twin transfusion syndrome:

Baldwin, V. J. 1994. *Pathology of Multiple Pregnancy*. Springer-Verlag, New York.

Keith, L. G., et al. (eds.), 1995. *Multiple Pregnancy*. Parthenon Publishing, New York.

Jacob, and Esau as donor and recipient MZ twins:

Baldwin, V. J. 1994. *Pathology of Multiple Pregnancy*, chapter 9, Springer-Verlag, New York.

The 90% rule for obligate brood reduction:

Simmons, R. E. 1988. Offspring quality and the evolution of Cainism. *Ibis* 130: 339–357.

CHAPTER 10. FAMILY HARMONY

Learning in osprey broods:

Edwards, T. C. 1989. Similarity in the development of foraging mechanics among sibling ospreys. *Condor* 91: 30–36.

Adaptive suicide:

McAllister, M. K., Roitberg, B. D. 1987. Adaptive suicidal behavior in pea aphids. *Nature* 328: 797–799.

McAllister, M. K., Roitberg, B. D., Weldon, K. L. 1990. Adaptive suicide in pea aphids: Decisions are cost sensitive. *Animal Behavior* 40: 167–175.

O'Connor, R. J. 1978. Brood reduction in birds: Selection for fratricide, infanticide, and suicide? *Animal Behavior* 26: 79–96.

Wilson, E. O. 1971. *The Insect Societies*. Belknap Press of Harvard University Press, Cambridge, MA.

Cannibalism in honeybees:

Kukuk, P. F., May, B. 1990. Diploid males in a primitively eusocial bee *Lasioglossum* (*Dialictus*) *zephyrum* (Hymenoptera; Halictidae). *Evolution* 44: 1522–1528.

Woyke, J. 1963. What happens to diploid drone larvae in a honeybee colony. *Journal of Apicultural Research* 2: 73–76.

Reciprocal altruism in vampire bats:

Wilkinson, G. S. 1984. Reciprocal food sharing in the vampire bat. *Nature* 308: 181–184.

Wilkinson, G. S. 1990. Food sharing in vampire bats. *Scientific American* (February): 76–8.

Polyembryony in wasps:

Beckage, N. E. 1997. The parasitic wasp's secret weapon. *Scientific American* (November): 82–87.

Cruz, Y. P. 1981. A sterile defender morph in a polyembryonic hymenopteran parasite. *Nature* 294: 446–447.

Grbic, M., Strand, M. R. 1992. Sibling rivalry and brood sex ratios in polyembryonic wasps. *Nature* 360: 254–256.

Strand, M. R., Grbic, M. 1997. The development and evolution of polyembryonic insects. *Current Topics in Developmental Biology* 35: 121–159.

Superparasitism in wasps:

Godfray, H.J.C. 1994. *Parasitoids: Behavioral and Evolutionary Ecology.* Monographs in Behavior and Ecology. Princeton University Press, Princeton, NJ.

Puttler, B., van den Bosch, R. 1959. Partial immunity of *Laphygma exigua* (*Huebner*) to the parasite *Hyposoter exiguae* (*Viereck*). *Journal of Economic Entomology* 52: 327–329.

van Alphen, J.J.M., Visser, M. E. 1990. Superparasitism as an adaptive strategy for insect parasitoids. *Annual Review of Entomology* 35: 59–79.

Primogeniture in humans:

Hrdy, S. B., Judge, D. S. 1993. Darwin and the puzzle of primogeniture. *Human Nature* 4: 1–45.

Pregnancy rates from IVF:

Schieve, et al., 1999. Live-birth rates and multiple-birth risk using in vitro fertilization. *Journal of the Aerican Medical Association* 282: 1832–1838.

Parent-offspring cooperation over incubation in gulls:

Evans, R. M., Wiebe, M. O., Lee, S. C., Bugden, S. C. 1995. Embryonic and parental preferences for incubation temperature in herring gulls: implications for parent-offspring conflict. *Behavioral Ecology and Sociobiology* 36: 17–23.

Born to rebel:

Sulloway, F. J. 1996. *Born to Rebel.* Pantheon Books, New York.

Dominant versus subordinate nestlings in blue-footed boobies:

Drummond, H., Canales, C. 1998. Dominance between booby nestlings involves winner and loser effects. *Animal Behavior* 55: 1669–1676.

CHAPTER 11. CANNIBALISM AND INFANTICIDE

Sexually selected infanticide:

van Schaik, C. P., Janson, C. H. 2002. *Infanticide by Males and Its Implications.* Cambridge University Press, New York.

Cannibalism in cardinalfish:

Okuda, N., Yanagisawa, Y. 1996. Filial cannibalism by mouthbrooding males of the cardinal fish, *Apogon doederleini*, in relation to their physical condition. *Environmental Biology of Fishes* 45: 397–404.

Okuda, N., Yanagisawa, Y. 1996. Filial cannibalism in a paternal mouthbrooding fish in relation to mate availability. *Animal Behavior* 52: 307–314.

Dracula ants and nondestructive cannibalism of larvae:

Masuko, K. 1986. Larval hemolymph feeding: A nondestructive parental cannibalism in the primitive ant *Amblyopone silvestrii Wheeler* (Hymenoptera: Formicidae). *Behavioral Ecology and Sociobiology* 19: 249–255.

Wheeler, G. C., Wheeler, J. 1988. An additional use for ant larvae (Hymenoptera: Formicidae). *Entomological News* 99: 23–24.

Infanticide in lions:

Bygott, D., et al. 1979. Male lions in large coalitions gain reproductive advantage. *Nature* 282: 839–841.

Heinsohn, R., Packer, C. 1995. Complex cooperative strategies in group-territorial African lions. *Science* 269: 1260–1262.

Packer, C., Pusey, A. 1983. Adaptations of female lions to infanticide by incoming males. *American Naturalist* 121:716–728.

Packer, C., Pusey, A., Eberly, L. 2001. Egalitarianism in female lions. *Science* 293: 690–693.

Pusey, A. E., Packer, C. 1994. Infanticide in lions: Consequences and counter-strategies. In Parmigiani, S., vom Saal, F. (eds.), *Infanticide and Parental Care*, pp. 277–299, Harwood Academic, London.

Sexual cannibalism:

Buskirk, R. E., et al. 1984. The natural selection of sexual cannibalism. *American Naturalist* 123: 612–625.

Elgar, M. A., Crespi, B. J. 1992. *Cannibalism: Ecology and Evolution among Diverse Taxa.* Oxford University Press, Oxford.

Gould, S. J., 1985. *The Flamingos Smile.* Norton, New York.

Parker, G. A. ,1979. Sexual selection and sexual conflict. In Blum, M. S., Blum, N. A. (eds.), *Sexual selection and reproductive competition in insects,* pp. 123–166. Academic Press, London.

Schneider,J. M., Elgar, M.A. 2001. Sexual cannibalism and sperm competition in the golden orb-web spider *Nephila plumipes* (*Araneoidea*): Female and male perspectives. *Behavioral Ecology* 12: 547–552.

Thornhill, R. 1976. Sexual selection and parental investment in insects. *American Naturalist* 110: 153-163.

Bruce effect:

Bruce, H. M. 1960. A block to pregnancy in the mouse caused by proximity of strange males. *Journal of Reproduction and Fertility* 1: 96–103.

Cinderella effect:

Daly, M., Wilson, M. 1988. Evolutionary social psychology and family homicide. *Science* 242: 519–524.

Daly, M., Wilson, M. 1999. *The Truth about Cinderella.* Yale University Press, New Haven, CT.

Daly, M., Wilson, M. 2001. An assessment of some proposed exceptions to the phenomena of nepotistic discrimination against stepchildren. *Annales Zoologici Fennici* 38: 287–296.

Wilson, M. I., Daly, M. 1994. The psychology of parenting in evolutionary perspective and the case of human filicide. In Parmigiami, S., vom Saal, F. S., (eds.), *Infanticide and Parental Care*, pp. 73–104. Harwood Academic Publishers, Chur, Switzerland.

CHAPTER 12. BRAVE NEW WORLDS

Hrdy, S. B. 1979. Infanticide among animals: A review, classification, and examination of the implications for the reproductive strategies of females. *Ethology and Sociobiology* 1: 13–40.

Embryo reduction of twelve live fetuses:

Lipitz, S., Frenkel, Y., Seidman, D. S., et al. 1994. Successful outcome of multifetal reduction in a pregnancy with 12 live fetuses. *Human Reproduction* 9: 1190–1191.

Infanticide in humans:

Ganatra, B. R., Hirve, S. S., Rao, V. N. 2001. Sex-selective abortion: Evidence from a community-based study in western India. *Asia-Pacific Popululation Journal/United Nations* 16: 109–124.

Haughton, J., Haughton, D. 1995. Son preference in Vietnam. *Studies in Family Planning* 26: 325–337.

Hausfater, G., Hrdy, S. B. 1984. *Infanticide: Comparative and Evolutionary Perspectives*. Aldine, New York.

Hrdy, S. B. 1997. *Mother Nature*. Pantheon Books, New York.

Junhong, C. 2001. Prenatal sex determination and sex selective abortion in central China. *Population and Development Review* 27: 259–281.

Kairys, S. W., Alexander, R. C., Block, R. W., et al. 2001. Distinguishing sudden infant death syndrome from child abuse fatalities. *Pediatrics* 107: 437–441.

Oomman, N., Ganatra, B. R. 2002. Sex selection: The systematic elimination of girls. *Reproductive Health Matters* 10: 184–197.

Parmigiani, S., vom Saal, F. S. (eds.). 1994. *Infanticide and Parental Care*. Harwood Academic Publishers, Chur, Switzerland.

Watts, C. Zimmerman, C. 2002. Violence against women: Global scope and magnitude. *Lancet* 359: 1232–1237.

Infanticide in twins:

Ball, H. L., Hill, C. M. 1996. Reevaluating "twin infanticide." *Current Anthropology* 37: 856–863.

Granzberg, G. 1973. Twin infanticide: A cross-cultural test of a materialistic explanation. *Ethos* 1: 405–412.

Rising incidence of twins:

Hogue, C. J. R. 2002. Successful assisted reproductive technology: The beauty of one. *Obstetrics and Gynecology* 100: 1017–1019.

Luke, B. 1994. The changing pattern of multiple births in the United States: Maternal and infant characteristics, 1973 and 1990. *Obstetrics and Gynecology* 84: 101–106.

Health risks of multiple gestations:

Bryan, E. M. 1986. The intrauterine hazards of twins. *Archives of Disease in Childhood* 61: 1044–1045.

The, ESHRE Capri Workshop Group. 2000. Multiple gestation pregnancy. *Human Reproduction* 15: 1856–1864.

Kovalevsky, G., Rinaudo, P., Coutifaris, C. 2003. Do assisted reproductive technologies cause adverse fetal outcomes? *Fertility and Sterility* 79: 1270–1272.

McCulloch, K. 1988. Neonatal problems in twins. *Clinics in Perinatology* 15: 141–158.

Powers, W. F., Kiely, J. L., Fowler, M. G. 1995. The role of birth weight, gestational age, race and other infant characteristics in twin intrauterine growth and infant mortality. In Keith, L. G., et al. (eds.), *Multiple Pregnancy: Epidemiology, Gestation and Perinatal Outcome*, pp. 163–174. Parthenon Publishing, New York.

Schieve, L., Meikle, S. F., Ferre, C., et al. 2002. Low and very low birth weight in infants conceived with the use of assisted reproductive technology. *New England Journal of Medicine* 346: 731–737.

Seoud, M.A.-F., Toner, J. P., Kruithoff, C. Muasher, S. J. 1992. Outcome of twin, triplet, and quadruplet in vitro fertilization pregnancies: The Norfolk experience. *Fertility and Sterility* 57: 825–834.

Cloning and imprinting:

Dean, W., Santos, F., Stopjkovic, M., et al. 2001. Conservation of methylation reprogramming in mammalian development: Aberrant reprogramming in cloned embryos. *Proceedings of the National Academy of Sciences* 98: 13734–13738.

Li, X., Li, Z., Jouneau, A., et al. 2003. Nuclear transfer: Progress and quandaries. *Reproductive Biology and Endocrinology* 1: 84–89.

Ogawa, H., et al. 2003. Disruption of imprinting in cloned mouse fetuses from embryonic stem cells. *Reproduction* 126: 549–557.

Reik, W., Dean, W., Walter, J. 2001. Epigenetic reprogramming in mammalian development. *Science* 293: 1089–1093.

Rideout, W. M. 3rd Eggan, K., Jaenisch, R. 2001. Nuclear cloning and epigenetic reprogramming of the genome. *Science* 293: 1093–1098.

Sinclair, K. D., et al. 2000. In-utero overgrowth in ruminants following embryo culture: Lessons from mice and a warning to men. *Human Reproduction* 15 (Supplement 5): 68–86.

Prenatal hormonal effects:

James, W. H., Orlebeke, J. F. 2002. Determinants of handedness in twins. *Laterality* 7: 301–307.

McFadden, D. 1993. A masculinizing effect on the auditory systems of human females having male co-twins. *Proceedings of the National Academy of Sciences* 90: 11900–11904.

Ryan, B. C., Vandenbergh, J. G. 2002. Intrauterine position effects. *Neuroscience and Biobehavioral Reviews* 26: 665–678.

vom Saal, F. 1989. Sexual differentiation in litter bearing mammals: Influence of sex of adjacent fetuses in utero. *Journal of Animal Science* 67: 1824–1840.

vom Saal, F., Bronson, F. 1980. Sexual characteristics of adult female mice are correlated with their blood testosterone levels during prenatal development. *Science* 208: 597–599.

vom Saal, F. S., Grant, W. M., et al. 1983. High fetal estrogen concentrations: Correlations with increased sexual activity and decreased aggression in male mice. *Science* 220: 1306–1309.

vom Saal, F., Vandenbergh, J. 1987. Regulation of puberty and its consequences on population dynamics of mice. *American Zoologist* 27: 891–898.

Risks associated with IVF and ART:

Bassil, S., Wyns, C., Toussaint-Demylle, D., et al. 1997. Predictive factors for multiple pregnancy in *in vitro* fertilization. *Journal of Reproductive Medicine* 42: 761–766.

Blickstein, I., Keith, L. G. 2001. The epidemic of multiple pregnancies. *Postgraduate Obstetrics and Gynecology* 21: 1–7.

Blickstein, I., Keith, L. G. 2002. Iatrogenic multiple pregnancy. *Seminars in Neonatology* 7: 169–176.

Elster, N. 2000. Less is more: The risks of multiple births. *Fertility and Sterility* 74: 617–623.

Jones, H. R. Jr. 2003. Multiple births: How are we doing? *Fertility and Sterility* 79: 17–21.

Lambert, R. D. 2003. Safety issues in assisted reproductive technology: Aetiology of health problems in singleton ART babies. *Human Reproduction* 18: 1987–1991.

Lucifero, D., Chaillet, J. R., Trasler, J. M. 2004. Potential significance of genomic imprinting defects for reproduction and assisted reproductive technology. *Human Reproduction Update* 10: 3–18.

Shinwell, E. S. 2002. Neonatal and long-term outcomes of very low birth weight infants from single and multiple pregnancies. *Seminars in Neonatology* 7: 203–209.

Templeton, A., Morris, J. K. 1998. Reducing the risk of multiple births by transfer of two embryos after in vitro fertilization. *New England Journal of Medicine* 339: 573–577.

Wilton, J. M. 1995. Breast feeding multiples. In Keith, L. G., et al. (eds.), *Multiple Pregnancy: Epidemiology, Gestation and Perinatal Outcome*, pp. 553–561. Parthenon Publishing, New York.

Progeny choice and infanticide:

Daly, M., Wilson, M. 1984. A sociobiological analysis of infanticide. In Hausfater, G., Hrdy, S. B. *Infanticide: Comparative and Evolutionary Perspectives*, pp. 487–502. Aldine, NY.

Edwards, M. L. 1996. The cultural context of deformity in the ancient Greek world. *Ancient History Bulletin* 10 (3–4): 79–92.

Hill, C. M., Ball, H. L. 1996. Abnormal births and other "ill omens": The adaptive case for infanticide. *Human Nature* 7: 381–401.

Hrdy, S. B. 1997. *Mother Nature*. Pantheon Books, New York.

Parmigiani, S., vom Saal, F. S. (eds.). 1994. *Infanticide and Parental Care*. Harwood Academic Publishers, Chur, Switzerland.

Zygote transfer:

Lundin, K., Bergh, C., Hardarson, T. 2001. Early embryo cleavage is a strong indicator of embryo quality in human IVF. *Human Reproduction* 16: 2652–2657.

Plachot, M., et al. 2000. Blastocyst stage transfer: The real benefits compared with early embryo transfer. *Human Reproduction Supplement* 6: 24–30.

Wittemer, C., et al. 2000. Zygote evaluation: An efficient tool for embryo selection. *Human Reproduction* 15: 2591–2597.

Folic acid supplementation, birth defects and twinning:

Erickson, A., Kallen, B., Aberg, A. 2001. Use of multivitamins and folic acid in early pregnancy and multiple births in Sweden. *Twin Research* 4: 63–66.

Green, N. S. 2002. Folic acid supplementation and prevention of birth defects. *Journal of Nutrition* 132: 2356S–2360S.

Lumley, J., et al. 2001. Modelling the potential impact of population-wide periconceptional folate/multivitamin supplementation on multiple births. *British Journal of Obstetrics and Gynaecology* 108: 937–42.

Index

Lightning Source UK Ltd.
Milton Keynes UK
UKHW021000211022
410853UK00006B/284